# Thinking by numbers

# 1

Written by SARAH KENNEDY

Series editor STEVE HIGGINS

**OXFORD**
UNIVERSITY PRESS

# OXFORD
## UNIVERSITY PRESS

Great Clarendon Street, Oxford OX2 6DP

Oxford University Press is a department of the University of Oxford.
It furthers the University's objective of excellence in research,
scholarship, and education by publishing worldwide in

Oxford   New York

Auckland   Cape Town   Dar es Salaam   Hong Kong   Karachi
Kuala Lumpur   Madrid   Melbourne   Mexico City   Nairobi
New Delhi   Shanghai   Taipei   Toronto

With offices in

Argentina   Austria   Brazil   Chile   Czech Republic   France   Greece
Guatemala   Hungary   Italy   Japan   South Korea   Poland   Portugal
Singapore   Switzerland   Thailand   Turkey   Ukraine   Vietnam

Oxford is a registered trade mark of Oxford University Press
in the UK and in certain other countries

Activity text © Sarah Kennedy 2005
Introduction text © Steve Higgins 2005

The moral rights of the author have been asserted

Database right Oxford University Press (maker)

First published 2005

British Library Cataloguing in Publication Data

Data available

ISBN-13 9780198361237
ISBN-10 019 836123 8

3 5 7 9 10 8 6 4 2

Illustrated by George Hollingworth
Typeset by Artistix, Thame, Oxon
Printed in Great Britain by Ashford Colour Press, Gosport, Hants

# Contents

# Introduction

*Thinking by Numbers* aims to develop thinking skills through mathematics lessons and activities across the primary age range. Although it can be used by an individual teacher, we think that you will get the best from the series if you use the activities across your school to undertake a professional inquiry into the potential of these lessons to develop pupils' thinking. Hence, the sections on *Professional development* (page 9), *Classroom management* (page 12), *Formative assessment and assessment for learning* (page 14), and *Speaking and listening* (page 18) are important aspects of the series. These sections will support you in helping to make the activities successful, as well as suggesting opportunities to develop aspects of your own teaching. Most of these introductory sections also contain suggestions for further reading that will support your exploration of thinking skills through the activities in *Thinking by Numbers*.

Teaching children to think for themselves is at the heart of primary education. It is all too easy to focus on the demands of the curriculum and its assessment and forget that the facts and knowledge have to be connected with an understanding of this curriculum content to help the learner make sense of it all. Without this understanding learners cannot use the information they have been taught and see how it relates to other ideas or knowledge that they have already. At the core of the thinking skills movement in education is the belief that this kind of thinking is teachable. This belief has been inspired by the work of two leading educators.

## History of thinking skills

In Israel after the Second World War, many refugee children had been through traumatic early experiences. On traditional tests, such as IQ tests or standardized tests of achievement, many of these children scored so badly that they seemed 'unteachable'. Working to integrate such children Reuven Feuerstein refused to accept this conclusion and devised ways to find out exactly which kinds of thinking they were unable to do, how they could be helped to develop these skills, and, therefore, each individual's *potential* for learning.

Feuerstein developed a set of techniques and tasks called 'instruments' that helped these learners succeed on subsequent tests. These methods were termed 'dynamic', in the sense that children were studying the process of learning and the change that took place. Feuerstein argued that such a process was much more likely to predict how a person might then learn in the future. Many of Feuerstein's ideas have influenced work on teaching thinking skills, in particular his emphasis on the importance of the interaction of the teacher, or 'mediation' of thinking.

Another important figure in thinking skills (or 'Critical Thinking', as it is called in the United States), is the American philosopher Matthew Lipman. As a university professor, he thought that his students had been encouraged to learn facts and to accept opinions, but not to think for themselves. He developed a programme, therefore, called 'Philosophy for Children', which aims to help younger people (from six-year-olds to teenagers) to think by raising questions about stories that they read together. The teacher uses children's natural curiosity about the stories in order to promote active participation and learning. One of Lipman's basic convictions is that children are natural philosophers, and that they view the world around them with curiosity and wonder, which can be used as a basis for thinking and reasoning.

Both Feuerstein and Lipman, though from very different starting-points, hold a similar belief in children's abilities. They have demonstrated that through thinking exercises and activities learners can exceed the predictions of achievement which tests may have suggested is their limit of competence. This, then, forms the basis of techniques in thinking skills – realizing children's potential. Their work has inspired many others to explore and develop approaches which help children to become more effective learners as they start to think for themselves. The aim of this book is to help you, as a teacher, to see how this kind of thinking can be developed.

## Teaching thinking

Some people argue that the idea of trying to teach general thinking skills is misguided because in practice thinking always occurs in a specific situation. Further, they believe that it is better to concentrate on teaching subjects and developing specific and detailed knowledge. However, *Thinking by Numbers* has been developed on the principle that there are common features of thinking in different situations, that it is helpful to try to apply techniques learned previously in new situations. For example, once you have used a graphic organizer, such as a Venn diagram, to compare and contrast themes in traditional tales in literacy, you can use the same technique to compare and contrast in other curriculum areas, such as family life in different eras in history.

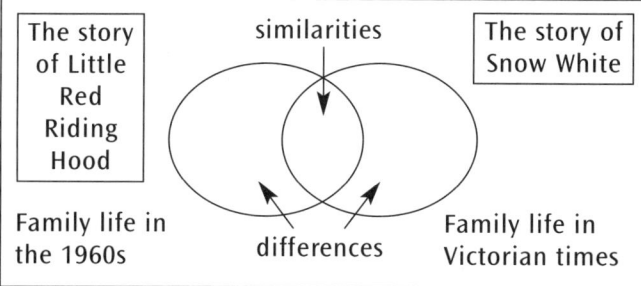

Since 1999, the national curricula for England and Wales now specifically include thinking skills (see page 6 for more details). In Scotland the *5–14 Guidelines* emphasize the capacity for independent thought through enquiry, problem solving, information handling and reasoning, as well as identifying learning and thinking skills in the core skills and capabilities. So the current challenge for teachers is not whether to teach thinking skills, but how best to teach them!

## Approaches to teaching thinking skills

There is a host of different programmes and approaches which advocate teaching thinking. These can be categorized broadly into whether they adopt an 'enrichment' approach where they are taught through extra or separate lessons, or an 'infusion' approach where the particular skills are taught through the normal lessons that schools provide. There are certain advantages and disadvantages to adopting these two different teaching approaches. If thinking skills are taught separately it is possible to make skills and techniques explicit, but there is a danger that they may not be used, except in special 'thinking' lessons. However, if they are taught as part of other lessons, such as mathematics or history, there is also a danger that the skills and techniques will become submerged by the curriculum content and not be seen as skills that can be applied elsewhere.

We believe that it is necessary to do both – to have a mixture of 'thinking lessons' with discussion of the kinds of thinking that are involved, and 'subject lessons' where skills can be applied and developed, but perhaps less explicitly. Identifying some lessons as 'thinking maths' lessons gives a clear signal to the children that you are looking for something different in the way that they work and the way they talk and listen. It is challenging to make the time to develop speculation or reasoning in every lesson, but it is also difficult to make sure it happens at *some* time in *some* lessons. We suggest that the activities in the different units can be used as a way to emphasize aspects of thinking that you wish to develop. You may then choose to develop other similar lessons where you can re-use the structure of the activities, or use some of the ideas and techniques in other subject areas.

*Suggestions for further reading*

H. Sharron and M. Coulter, *Changing Children's Minds: Feuerstein's Revolution in the teaching of Intelligence* (Birmingham, Questions Publishing Company, 1994)

M. Lipman, *Thinking in Education* (Cambridge University Press, 2003)

C. McGuinness, *From thinking skills to thinking classrooms: A review and evaluation of approaches for developing pupils' thinking* [DfEE Research Report RR115] (Norwich, HMSO, 1999)

V. Wilson, *Can Thinking Skills Be Taught? A paper for discussion* (Edinburgh, Scottish Council for Research in Education, 2000) [Available at: http://www.scre.ac.uk/scot-research/thinking/index.html]

# Thinking skills and the National Curriculum

## Classifying thinking skills

There are more ways to think about thinking than you could imagine! Amongst the wealth of lists, frameworks, models and taxonomies of thinking that have been developed, many people have heard of the 'Bloom's Taxonomy', which is considered the original way of classifying 'higher order thinking'. This taxonomy is basically a three-tier model:

- **knowledge** – in the form of facts, concepts, rules or skills
- **basic thinking** – relatively simple ways of understanding, elaborating and using what is known
- **higher order thinking** – a learning process which leads to a deeper understanding of the nature, justification, implications and value of what is known.

The National Curriculum in England uses five classification headings to denote thinking skills that should be embedded across all subject areas so pupils learn how to learn. These are:

- evaluation
- creativity
- enquiry
- reasoning
- information processing.

However, no single classification or framework can ever fully describe the complexity of all the kinds of thinking we experience. What is missing in both the original version of Bloom's work and in the National Curriculum list is the role of the thinker in thinking – one's own awareness, reflection and engagement. This metacognitive component (i.e. thinking about how we think) is an essential ingredient in developing a learner's understanding of their own thinking and the ability to think for oneself. The following table shows how *Thinking by Numbers* works alongside these thinking classifications to develop thinking skills.

| Bloom's Taxonomy | National Curriculum | Thinking by Numbers | |
|---|---|---|---|
| **Knowledge**<br>Abstracts and universals<br>Using specifics<br>Knowledge of specifics | *Information processing* | Unit 1: Sort it out! | *Unit 6: Think on!* |
| **Basic thinking**<br>Application<br>Comprehension | | | |
| **Higher order thinking**<br>Evaluation<br>Synthesis<br>Analysis | *Reasoning*<br>*Enquiry*<br>*Creativity*<br>*Evaluation* | Unit 2: That's because …<br>Unit 3: Detective work<br>Unit 4: What if?<br>Unit 5: In my opinion | |

Comparison of Bloom's Taxonomy, the National Curriculum and *Thinking by Numbers*

## The National Curriculum

The National Curriculum categories contain the following breakdown of skills, which form the basis for the units in the *Thinking by Numbers* series.

### Information processing skills

These enable pupils to locate and collect relevant information, to sort, classify, sequence, compare and contrast, and to analyse part/whole relationships.

### Reasoning skills

These enable pupils to give reasons for opinions and actions, to draw inferences and make deductions, to use precise language to explain what they think, and to make judgements and decisions informed by reasons or evidence.

### Enquiry skills

These enable pupils to ask relevant questions, to pose and define problems, to plan what to do and how to research, to predict outcomes and anticipate consequences, and to test conclusions and improve ideas.

### Creative thinking skills

These enable pupils to generate and extend ideas, to suggest hypotheses, to apply imagination, and to look for alternative innovative outcomes.

### Evaluation skills

These enable pupils to evaluate information, to judge the value of what they read, hear and do, to develop criteria for judging the value of their own and others' work or ideas, and to have confidence in their judgements.

The first five units of the *Thinking by Numbers* books are based on these classifications. We have also added two further components:

- ◐ a final unit which provides opportunities for **using and applying thinking skills** covered in the earlier units
- ◐ a **metacognitive skills** element running throughout all of the activities, which aims to develop children's awareness and understanding of the thinking they are doing.

> *Suggestions for further reading*
> L.W. Anderson and D.R. Krathwohl (eds.),
> *A Taxonomy for Learning, Teaching and Assessing:*
> *A revision of Bloom's Taxonomy of Educational*
> *Objectives* (New York, Longman, 2001)
>
> S. Higgins, J. Miller, D. Moseley and J. Elliot,
> 'Taxonomy Heaven', *Teaching Thinking*, 12,
> (Autumn 2003)

# Thinking skills in mathematics

Mathematics is an area of the curriculum which is full of opportunities to develop pupils' thinking skills and reasoning abilities. An emphasis on developing strategies, identifying patterns and rules, and clarifying concepts helps children learn mathematics by making aspects of it more explicit in the classroom. Developing reasoning, problem solving and enquiry skills through mathematics can support the development of these 'higher order' thinking skills more widely, and encourage successful learning in other subjects. A number of principles underpin the activities in each of the six units in *Thinking by Numbers*. These will help pupils to see the connections between the way that they have worked on a mathematics task and then how they can apply these skills in other contexts, either in other areas of mathematics or other areas of learning and understanding.

## Challenge

Thinking activities must provide a level of challenge. This means that they should not be too easy to complete, nor so hard that the pupils cannot recognize that the activities have been successful. Alternatively, the activities may have more than one solution, or route to a solution, that can be evaluated by the pupils to decide which is the best answer or approach. Mathematics is a subject which people often think they just 'can't do'. Successfully completing challenges encourages pupils to see that maths is a subject that they can learn to be good at.

## Active discussion

Thinking activities need to be talked about. Mathematics has both vocabulary and a language of its own. Familiar words are used in unfamiliar ways, such as 'product' or 'difference', and it has its own terminology, such as 'numerator' or 'perpendicular'. Pupils will need time to practise speaking mathematically and explain what they are thinking using this language. This can be difficult to do with the whole class, so some paired or small group work is essential to provide opportunities to explore ideas and allow pupils to develop confidence with the vocabulary.

## Feedback

Giving feedback will be key in ensuring pupils make progress in a thinking activity. One of the easiest ways to do this is to have 'mini-plenaries' as the lesson develops. Stop the class for a few minutes and ask a group to explain where they are up to. This will give you the chance to highlight successful ways of working, as well as asking for reasons and challenging their thinking.

## Review

When developing thinking skills it is important to review both the **content** of the activity and the **process** that the pupils have used to complete the activity. This means talking about the mathematics involved in the task and the way that they have worked (the skills used in collaborating, working systematically, or identifying patterns and rules). It is often helpful to discuss the latter the next time pupils undertake a similar task so that you can remind them of what was successful. A combination of 'mini-plenaries' throughout the lesson, a review at the end of a lesson, then recapping at the beginning of the next lesson will help ensure children understand that you want them to think not just about *what* they have learned, but *how* they have learned it.

# Professional development

We advocate that you try out the *Thinking by Numbers* activities as part of your professional development programme. A critical perspective on the lesson is essential. The activities alone will not succeed in developing thinking skills without this perspective. It is helpful to have a colleague with whom to discuss the activities as you try out the different ideas. We believe that a key part of teaching thinking and thinking skills successfully is to have some time and space to reflect on your own teaching so as to increase the emphasis on developing pupils' understanding. The introductory 'brief' and final 'debrief' sections of each activity aim to support this by summarizing the key features of the lessons and indicating aspects for review.

*Thinking by Numbers* provides a combination of teacher-led activities, then discussion and collaborative working in small groups, followed by some kind of whole class discussion, or plenary, which reviews both the content and process of learning. The results of this approach are usually a higher level of engagement in the activities, more talking and discussion about the activities. The activities themselves are open-ended to the extent that genuine discussion is not only possible but helpful. They are also challenging but enjoyable activities, helping to create a classroom climate where there is an emphasis on succeeding after effort.

As part of this process you should get more opportunities to hear what your pupils think. As you plan these lessons to increase engagement in learning you will need to listen carefully to how your pupils respond. The enjoyment should initially help to sustain more permanent changes in patterns of classroom interaction. The further feedback you get from insights into pupils' understanding will help identify any misunderstanding or misconceptions that you can tackle through 'mediation' or questioning and discussion.

Some suggestions for getting started:

1   **Work with a colleague**. This might be a colleague teaching the same year group, in which case you can investigate the impact of the same activities. Alternatively you may be working with a colleague in another year group, so you might look at similar kinds of activities or similar aspects of thinking. Working with a colleague means you are more likely to keep to your plan, building progress in time for review. Discussing things with someone else helps to clarify our own thinking, and makes it easier to see patterns or themes in what has happened.

2   **Decide what you want to investigate or improve**. It is easier to develop children's thinking if you focus on a particular area that you feel needs improvement. You could:
    - identify information processing as a key mathematical skill needing improvement
    - focus on your own questioning and how you probe and challenge your children's thinking
    - develop more precise use of mathematical language
    - aim to increase participation in lessons by children who are not usually engaged.

3   **Set a timescale** (at least eight weeks, up to a school year) and plan which activities you are going to use. How often will you have *Thinking by Numbers* lessons? Once a week? Once a fortnight? How will you make sure you have time to review the activities with a colleague?

4   **Try out the activities and review** them as soon afterwards as you can with your colleague. What was different in the lesson compared with other maths lessons? Were you able to see patterns in the children's thinking? Were there any common misconceptions that you needed to tackle? How well did the collaborative tasks go?

5   **Analyse what happened**. If there is improvement, what do you think caused it? The focused practice? Your extra time and effort? The pupils' discussion? Your understanding of their thinking? Would it probably have happened anyway?

6   **Review progress**. What have you learned that you can apply in the longer term? Do some kinds of questions work better than others? Can you use any of the strategies more widely?

# How to use *Thinking by Numbers*

The activities in *Thinking by Numbers* can be used in different ways – there is no need to work through them in order, though the final unit is designed to let pupils apply the skills that they have developed. Therefore, for you to assess how well these skills have been learned, it should be used after some of the other activities in the first five units. Some of the activities are based on thinking skills strategies which can be used more widely either in mathematics or other subjects of the curriculum. You should therefore evaluate if there are any aspects of the activity or teaching technique which could be used more generally. Although the books are aimed at different age groups, you may find activities that you can use or adapt in other books in the series. This is particularly true of the generic strategies, such as 'odd one out', which can be used again in mathematics or other areas of the curriculum. See page 24 for a fuller description of some generic activities used throughout the series.

Just as enquiry is at the heart of thinking skills activities for pupils, we believe that it also needs to be a part of the way you use them as a teacher. None of the activities will work by themselves, and they will not all be equally effective since this depends on the existing skills and knowledge of your pupils. You will have to use them critically to see how they can help your pupils' thinking – it is impossible to do this directly, since we cannot see into our pupils' heads and know what they are thinking. Nevertheless, it is possible to plan a series of activities that enable you to find out about pupils' thinking at different times and in different ways. This allows you to infer their level of understanding. Therefore, this needs to be a process of enquiry – finding out what and how your pupils think. The 'Watch out for' and 'Listen for' sections of each activity should help with this process.

## The units

The units are based around the classification of thinking skills in the National Curriculum for England and the headings of **information processing**, **reasoning**, **enquiry**, **creative thinking** and **evaluation**. Each unit begins with an overview of these particular aspects of thinking, and ends with a summary looking at how these skills can be developed. Of course, it is not possible to separate the thinking in different activities so that they only involve reasoning or creative thinking. Thinking is a complex activity which involves all kinds of thinking at the same time. It is holistic, multi-dimensional and dependent upon the context that we find ourselves in. The purpose of the tasks in *Thinking by Numbers* is to enable you to focus on a particular kind of thinking and to consider how it can be developed or fostered in your pupils.

## Links

The appendices (pages 88 to 95) contain information about how *Thinking by Numbers* relates to the *Framework for Teaching Mathematics* used in England, and the *5–14 Guidelines* for Scotland. A glossary of thinking skills terms is also included on page 96 for reference.

---

### Suggestions for further reading

P. Adey, *The Professional Development of Teachers: Practice and Theory* (Dordrecht, Kluwer Wolters, 2004)

S. Higgins, *Thinking Through Primary Teaching* (Cambridge, Chris Kington Publishing, 2001)

## The activities

**❹** Each activity has a whole class introduction where you will be 'Setting the Scene' and modelling the problem to the children.

**❶** Each activity has an introduction, 'The Brief', and review points, 'The Debrief', to explain the context for the activity. This is the 'professional development' part to help you consider what you want to achieve in the thinking lesson, and later to review how well it achieved its thinking skills aims.

**❻** The 'Checkpoints' section gives ideas for how to the keep the activity on track. This section also has suggestions on what to watch and listen out for, and prompts and pointers to stimulate discussion.

### Journey to the Moon

BRIEF

In 'Journey to the Moon' the children solve the problem of getting to the Moon. It's a game that involves exploring sequences of numbers, finding combinations and developing systematic approaches to recording for the different moves. While the game format helps all the children participate, it gives you the opportunity to hear their thinking as they justify their ideas. The task allows you the opportunity to observe and develop their reasoning skills through questioning.

**Key maths links**
- ○ Counting, properties of number and number sequences
- ○ Place value and ordering
- ○ Checking results of calculations

**Thinking skills**
- ○ Giving reasons
- ○ Working systematically
- ○ Describing

**Language**
count on, adding, altogether, one more, two more, total, more, equals What could I try next? I think ...

**Resources**
PCM 5 (one per pupil and one enlarged)
PCM 6 (one per pupil)
copy of *Man on the Moon: A day in the life of Bob* by Simon Bartram or non-fiction books on the Moon and travelling through space
1–6 dice (one per pair)
counters

**① Setting the scene**

The children play a version of 'Snakes and ladders' where they throw the dice to progress but have to write down the addition calculation correctly before they can move.

**② Getting started**

Use an enlarged copy of PCM 5 and demonstrate how to play the game. *Are you all clear how to play? What happens when you land on the end of the rocket jet?* (move up to the top of the rocket) *What happens when you land on the star?* (move down to the bottom of the star) *Where is the first rocket boost? If I am on square 3 and I want to get to 6 what number would I want to roll?* Take responses and agree that 3 more steps would be needed. Encourage the children to play the game a few times. In pairs, using PCM 5 and a dice, ask the children which pairs of numbers would get them from 3 to 9 in two moves. (model 1 + 5 = 6, 2 + 4 = 6 etc.)

**Simplify**

In pairs, using PCM 6 and a dice ask them which pairs of numbers would get them from the start to 6 in 2 moves. (model 3 + 3 = 6) *What is the biggest number you could use? What is the smallest number?* Continue until all the numbers are recorded. Help the children to record their moves as a number sentence. *Do the number sentences help?*

**Challenge**

*Your counter is on 8. You roll a dice and after two moves you land on 15. Can you find all the ways to get to 15?* The children will need to work in pairs. They can use PCM 6 to aid their recordings. They can use arrows on the number ladders to make number sentences. The number ladders can be used for different starting points.

**③ Checkpoints**

Are the children recording their progress as number sentences? Look for examples that show methodical working. This is a good opportunity to model different ways of working and invite children who are working systematically to share their methods. How well have they engaged in the task? Was the pairing successful?

How well are they counting on? This is a good opportunity to model the language of adding two scores together.

**! Watch out for ...**

Some children may count the square that they are on rather than starting with the next square. Model the method of moving on. Prompt the children to check for repeats, e.g. 2 + 4 = 6 is the same as 4 + 2 = 6, despite the order of the dice throw. Check that they know when all possibilities have been found. Encourage the use of systematic approaches. Are they clear using the language of addition? Can they record their thinking clearly?

**? Ask ...**
- ○ Can you give a reason?
- ○ What if ...?
- ○ What could I try next?
- ○ How do you know?

**" Listen for ...**

Observe the children as they talk to each other. Listen for children using appropriate vocabulary and ways of recording addition, e.g. number lines.

**⚡ Moving on ...**

Investigate pairs of numbers to find all of the possible combinations for different totals. Write a class list in a systematic order to model reasoning. Discuss if there were gaps or repeats. Then, complete the list filling the gaps and deleting the repeats.

**Where next?**
- ○ Play the game fully and then ask the children to design their own versions.
- ○ Investigate totals when throwing dice. Which numbers come up most often?
- ○ Use a table to show addition totals.

What worked well? How did the children work as pairs? Were they able to take turns when playing the game? Do you need to do more work on the mathematical objectives? Were there any surprises?

DEBRIEF

36 / 37

**❷** Basic information about mathematical objectives and language are included, along with any resources needed, plus the thinking skills focus.

**❺** The 'Getting started' section shows how the activity can be developed through collaborative group work.

**❼** Reviewing progress and stimulating further thinking are covered in the 'Moving on' section, as are suggestions to develop the teaching strategy or approach in other mathematics lessons, or in other subjects in 'Where next?'.

**❸** Each activity has accompanying photocopiable resources. Some are resource sheets for the activities, others aim to support the recording of the activities, particularly by pairs or small groups of pupils (for more information on recording see page 16).

# Classroom management

## Structure and timing of lessons

Each book in the *Thinking by Numbers* series comprises six units. These are based around the English National Curriculum thinking skills headings of **information processing, reasoning, enquiry, creative thinking** and **evaluation**, with a final unit focused on using and applying the skills acquired through the earlier activities. Each unit contains two activities, with three in the final 'using and applying' unit, giving a total of 13 thinking activities or lessons for each year group. They have been planned as mathematics lessons and cover aspects of the curriculum appropriate for each age group (see the NNS and *Mathematics 5–14* matching charts on pages 90 to 95). You could also use the activities as thinking lessons and follow the suggestions at the end of each unit to develop the ideas and thinking strategies across the curriculum. When planning how to use the activities there are a number of different approaches you could take, and these are outlined here.

### Regular thinking skills development

You could work through the activities using a *Thinking by Numbers* activity every two or three weeks. The benefit of this approach is that it provides regular opportunities to highlight the thinking skills you want to develop across the year. In the intervening time you would need to make sure that you refer back to the lessons and activities, as it would be all too easy for your pupils, particularly younger children, to forget what you are looking for in their work in the thinking activities.

### Intensive thinking skills development

You could choose to work through the units more intensively, perhaps one activity each week over a term, so that you could then take the skills and ideas further over the course of the year. This may also be more suitable for year groups in England where your teaching is affected by statutory tests, such as Year 6 in particular. A further advantage of this approach is that you can build up some momentum with regular 'thinking maths' lessons. Assuming they go well initially, the children will start to look forward to the lessons and you can then capitalize on this enthusiasm. You will also

develop a language around the lessons and activities with your class, and the regular practice will enhance this development.

### Integrated thinking skills development

Another possible approach for teachers in England and Scotland is to use the matching charts to the *NNS Framework* or *Mathematics 5–14*. These are provided in the appendices and will enable you to substitute the *Thinking by Numbers* activities where they fit most appropriately in your usual teaching plan. Whilst this is less disruptive to the mathematics curriculum, you will need to work hard to develop the thinking themes in the book. The thinking skills issue here is how you get the children to use what they learn elsewhere. This is always a challenge with any learning at school: how do you get learners to transfer what they know or can do to a new situation? The concept of 'bridging' is a useful one. As a teacher you connect or 'bridge' the knowledge or skills between different contexts. Where you have regular lessons you can mention things that you then refer to in other lessons. The further apart the sessions, the harder you will have to work to make those connections meaningful. *You remember when we used a Venn diagram to look at similarities and differences? Could we do something similar here?*

## Managing the lesson

To use the *Thinking by Numbers* activities effectively you will need to think through the method of working. Your pupils will need to have a clear idea of what they are doing, and why, so that in the review sections of the lesson they can evaluate how successful they have been. It is important to get the lessons off to a good start, so the children will need a 'hook' or some initial stimulus to launch into the activity well. This can be either through the way you introduce the activity, the resources that are used, or perhaps the way you make it meaningful to the pupils, tapping into their particular interests or enthusiasms. It is hard to predict exactly how long the different activities will take to complete. Sometimes children become particularly enthusiastic about a particular task and you will struggle to get through everything that is suggested. On other occasions you will have time to review the activities and ask the children to reflect on their learning.

## Introducing the lesson

Each activity begins with some kind of whole class introduction or demonstration. In this part of the lesson it is important to explain the activity and its purpose clearly. You should make objectives explicit; explain what you want from the pupils in terms of how they should work and the kind of language they should use. You will need to get feedback from the pupils to evaluate whether they understand what they are doing and know what they will have to do in the next phase of the activity. You may also need to adapt the activities according to the needs of your pupils. Although the activities have been designed for particular ages of pupils, you will need to judge whether some alteration is needed to provide the appropriate level of challenge for your class.

## During the lesson

In most of the activities the pupils apply or extend the ideas presented in the introduction by working collaboratively in pairs or small groups. When moving from whole-class to paired or group work, it is useful to discuss or mention how the pairs or groups are going to work together and what you are looking for. At the transition it is worth praising specific behaviours: *I liked the way you sorted out the number cards for your group, David.* Though it is also important to tailor this praise, particularly for older pupils who should be aware of supportive behaviours and active listening strategies: *Your group got started really quickly, Emma, what was it that you each did?* Reinforce the method of sharing ideas, explaining that they can do better together than they can separately, and that copying and ownership of ideas are not factors. The tasks themselves are designed to be challenging and to benefit from some discussion in small groups so that pupils don't just make up their minds quickly. The activities also contain suggestions for differentiation, with advice on simplifications and challenges that should help you to ensure that the level of challenge is maintained as the pupils work through the tasks. Further advice on

the opportunities to develop speaking and listening skills are outlined on pages 18 and 19.

## Reviewing the lesson

The hardest part of these activities is in helping pupils to see that particular tactics, strategies or approaches are helpful, without teaching specific solutions or answers. This will require some skilful questioning and discussion. It is important to review both the process that the pupils have used, particularly the collaborative skills of speaking and listening, as well as reviewing the curriculum content and knowledge and understanding of the activities.

It is also a good idea to review some of this as the lesson unfolds, rather than waiting until the end. Whilst the plenary seems to be the logical place to review the lesson, the pupils also know that the lesson is drawing to an end and it can be hard to maintain their interest. Mini-plenaries are, therefore, an essential teaching strategy which can help make the activities successful. These can be very brief, just checking where groups are up to, or sharing a successful technique or tactic being used by some children. *I noticed you've sorted the cards into different groups, can you tell the class how they are organized?* It boosts their confidence if you draw this to the attention of the whole class and gives other pupils who may not be on track a clear hint about what they could do.

Another possibility is to recap at the beginning of the next lesson. This is essential if the *Thinking by Numbers* sessions are a week or more apart. You need to remind the children that these are different lessons which require thinking, explaining, reasoning and evaluating. There should be more time for discussion about what went well previously and what skills or strategies they might find useful. The main aim is to help the pupils understand that they might not be able to see a solution immediately, but by thinking and working together they will be able to complete the activity successfully. In mathematics this is particularly important as it is a subject which pupils tend to think that they are either good at or not good at, rather than a subject that they can all learn to be better at!

# Formative assessment and assessment for learning

Formative assessment is about intervening during teaching to improve learning. As a teacher you gather feedback about what is going on (either within a lesson or between lessons) and use that information to alter what you do subsequently. Assessment for learning is a more interactive approach that takes assessment a stage further by involving the learners in understanding what the specific learning objectives are for each activity/ task/lesson so that they can judge how successful they have been in achieving them. This helps teachers and pupils to understand the criteria for being successful at learning, both for short term objectives as well as longer term goals about 'learning to learn' more effectively.

When using assessing for learning it is important to give pupils feedback about what they can do to improve (rather than giving marks or feedback that simply indicates whether they are correct or not). One common technique is to get pupils to give you feedback about how well they think they are doing on an activity or a piece of work. This can be a simple thumbs up/down signal from the class, or getting pupils to use traffic light colours to self-assess a piece of work they have done – green for go ('I understand it and can go on'), orange for getting there ('I could do with a little bit of help'), red for stop ('I'm stuck').

Thinking skills approaches also involve formative assessment. Most of the activities are about giving you, the teacher, information about children's thinking. This lets you assess their understanding and make decisions about how to support the development of that thinking. In addition, pupils are expected to talk about their thinking as they undertake the tasks. Developing this metacognitive talk (talk about their own thinking) is a powerful technique which helps learners understand their learning better.

Furthermore, focusing on what makes for successful learning encourages judgement about that learning and moves the discussion away from the products

or outputs (such as a complete page of calculations) to what has been learned (such as, 'I am finding subtraction more difficult than addition'). The concept of transfer is crucial here since it moves learning away from the particular to the more general. *What have you learned today that you can use in the future? What have you learned previously that will help you now?*

Both assessment for learning and thinking skills approaches use collaborative techniques for learning: paired and group work so that learners benefit from discussion with their peers. Both approaches highlight the role of the teacher in effective questioning and discussions with the pupils to move their thinking on. Assessment for learning and thinking skills approaches are clearly complementary. If you are developing formative assessment you will be developing children's thinking skills. If you are developing children's thinking skills and being explicit about the thinking they are doing with them, then this is formative assessment!

## Suggestions for further reading

Primary National Strategy, *Excellence and Enjoyment: learning and teaching in the primary years. Planning and assessment for learning: assessment for learning* (Document code: DfES 0521-2004 G) (2004)

Assessment Reform Group, *Assessment for Learning: 10 principles* (London, QCA, 2002) (available online at: http://www.qca.org.uk/ages3-14/ downloads/afl_principles.pdf )

P. Black, C. Harrison, C. Lee, B. Marshall and D. William, *Assessment for Learning. Putting it into practice* (Maidenhead, Open University Press, 2003)

S. Clarke, *Unlocking Formative Assessment: Practical strategies for enhancing pupils' learning in the primary classroom* (London, Hodder and Stoughton, 2001)

# How do you know it is working?

One of the greatest challenges in developing learners' thinking is assessing how well the activities are going. You should feel that the tasks and activities are giving the children opportunities to think and you should get direct and indirect evidence of this. There are a number of ways that you can start to gauge the impact of the activities.

## Enjoyment

First and foremost the activities should be enjoyable, both for you and your class. It is important that the activities are regarded as fun because this helps the children to develop their confidence to discuss what they think. It encourages the children to offer opinions and ideas without the worry of being 'wrong'. This aspect of the activities is vital to ensure their success. Thinking is hard work, so it needs to be as enjoyable as possible!

## Participation

Enjoyment should lead to increased engagement and involvement in the lessons. One of the ways that you can assess this is by keeping track of who participates. Are the contributions coming from those who are usually involved and usually speak in whole class discussions? Can you use the paired or group work to build pupils' confidence in contributing to a whole class discussion? *I thought that your suggestion was a really good one – can you explain it to the class?* Are you getting spontaneous contributions from those you normally have to ask directly?

## Language

The next thing to watch for is language that indicates thinking and reasoning. Are the pupils giving reasons? Do they use words like *then, so, because*? Are they being tentative (*I think … It could be … It might be …*) or speculative (*What if …? How about if we …?*)? You can start the lesson by saying you want to hear particular phrases, and giving suggestions for how they may be used. Then you need to look out for these first when the children are working in pairs or small groups. Then encourage the children to give longer responses in class discussions, ask them for reasons or examples, or to comment on each other's ideas. One of the most effective ways of encouraging this is simply to wait longer when you ask a question, and wait a little bit longer at the end of the response whilst indicating that you want them to continue. In mathematics you should also see the children using specific vocabulary more precisely; for example, are they getting more accurate in the use of words like *number, numeral* and *digit*? Or terms like *side, corner, edge* and *vertex*? You should also pay attention to the questions that the children ask. If the lessons are successful, the children will be asking questions about the content of the learning (rather than just about what they have to do).

## Reflection

If the activities are working the children should know that they have been successful and that they have been thinking hard. They should show growing awareness of this and come out to talk about their thinking. At first this will come out during the activities or just as you finish. It is a good tactic to get them to review and reflect at the beginning of the next *Thinking by Numbers* task; this will help remind them of what is expected in the next task as well as giving you a chance to assess how much they recall from last time!

## Transfer

The long term goal of *Thinking by Numbers* is to develop transferable skills. Evidence of this is shown when children start to refer back to thinking skills activities in terms of what they have learned. You should, therefore, begin to notice that they are using and talking about the skills that they are developing in other maths lessons or in other subjects. If this is spontaneous or unprompted you know that they are using the thinking skills for themselves.

# Recording

Opportunities for recording are identified in most of the activities. However, there are a number of issues you will need to consider. The activities are about developing thinking and the lesson must focus on this as the most important outcome. Recording can distract from this if the children become concerned with making sure they 'get it right' when they have to write something down. There are two main aspects of recording. The first is the recording of the particular task. Some of the photocopiable resources are explicitly designed for this. For other activities the children will need to think about the best way to record their thinking and their progress through the activity. The activities are often collaborative so you may need to make copies of the completed sheet for all the children in the group.

The second aspect of recording is to support review of the activities. The 'What did you learn today?' photocopiable sheet (see page 17) is designed to help with this. It may not be appropriate to use it for every activity, but it will help you review aspects of the lesson that enable the children to develop an understanding of their thinking and their learning (see *Formative assessment and assessment for learning* on page 14 for more information about developing thinking about learning). This aspect is cumulative and progressive as you will need to encourage the children to think about:

- their learning
- what they did
- what kind of thinking was involved
- how they worked together
- what lessons or skills they have learned that they can use in the future.

When planning how to incorporate recording into a thinking lesson, it is helpful to consider the following principles.

### 1 Recording should be purposeful
The record should either help with the process of the task or capture aspects of the thinking that it will be helpful to review.

### 2 Recording should be integral
If keeping track of what they are doing is not part of the task, it becomes an extra burden and less likely to be completed effectively.

### 3 Recording should be used
If you ask the children to make some notes on their thinking, or to use the 'What did you learn today?' PCM, you need to make use of it in a discussion either in that lesson or as part of setting the scene for the next activity.

### 4 Recording should be short
The lessons are about thinking and this needs to be the most important part of the lesson. You will not be able to capture everything that happens; you may need to have some kind of record to keep track of what has happened, but keep it as simple as possible.

# What did you learn today?

Name _____ Date _____

What did you learn today? _____

_____

_____

## What kind of thinking did you do today?

| | Yes | No |
|---|---|---|
| I remembered things to help me | ☐ | ☐ |
| I sorted things to help me | ☐ | ☐ |
| I used words like 'because' and 'then' | ☐ | ☐ |
| I found out something new | ☐ | ☐ |
| I used a rule to work something out | ☐ | ☐ |
| I had a new idea | ☐ | ☐ |

## How challenging was it?

Circle one of the choices on the line.

very easy        easy        OK        hard        very hard

## Working with others

| | Yes | No |
|---|---|---|
| I asked my teacher a question | ☐ | ☐ |
| I asked my partner a question | ☐ | ☐ |
| I shared my ideas | ☐ | ☐ |
| I was good at listening to my partner | ☐ | ☐ |

# Speaking and listening

Talking, thinking and learning are all closely related. We can remember things that we have heard, but it is only when we can put these ideas into our own words that we know we have learned them effectively. Speaking and listening are, therefore, at the heart of any thinking skills work. Listening to your pupils talk is also the best feedback you can get to assess what they are actually learning. It is therefore essential that the lessons and activities have speaking and listening at their core.

Children should be able to explain not just what they are doing, but why, and that their thinking is about the learning they are involved in. This involves speaking, listening and participating effectively in small and large group discussions. This helps them to learn by using new vocabulary (or words they already know more accurately) to express new ideas and new thinking. This process is difficult and requires time and support. Part of the purpose of the group work is to allow this to happen. Children will hear their peers making suggestions and having ideas about the tasks. As they join in and make their own suggestions they will work together to find a solution. This will help children succeed more independently in future tasks. The discussions with the whole class will help them to be more confident in what they are saying and thinking, and will give you opportunities to provide feedback on what you are looking for in thinking lessons. The table on page 19 sets out a progression in speaking, listening and group discussion and interaction across the primary age range.

## Classroom language

Classroom language is like a dialect of English. It has particular features and implicit rules that are different from language outside of school. The way you take turns, as a pupil, is very different from the way you normally take turns in conversation, either with your friends or at home. The teacher's use of questions, in particular, is strikingly different. Questions are often heavily loaded. For example, if you ask 'Why did you write that?', a child may assume that you are challenging them because it is incorrect and that they should have put something else. In a thinking skills lesson you may be wanting them to explain the reasons for their choices, or the decisions they made about what to write down, so as to provide a model for the rest of the class. If a teacher asks 'What do you *think* you should do?', the pupils may assume that you are reprimanding them for not listening, rather than asking them to speculate. It is therefore very important to think carefully about the questions that you ask to try to ensure that your pupils understand you really *do* want to know what they are thinking! Some examples of good questions are provided on page 21.

## Talking maths

Mathematical language is also different from everyday English. It is important that children do not just learn and remember the vocabulary, but learn how to use the language to communicate. This will help them to develop their mathematical thinking. Many words have specialist meanings in maths lessons, such as 'odd' and 'even'. Other words may not be encountered outside of these lessons, for example, 'trapezium' and 'numerator'. The *Thinking by Numbers* activities are a chance for children to speak the language of mathematics, rather than just practise its vocabulary.

**Suggestions for further reading**
Primary National Strategy, *Speaking, Listening, Learning: Working with children in Key Stages 1 and 2. Professional development materials* (Document code: DfES 0163-2004) (2004)

N. Mercer, *Words and Minds: How We Use Language To Think Together* (London, Routledge, 2000)

S. Higgins, *Parlez-vous mathematics? Enhancing Primary Mathematics Teaching and Learning,* I. Thompson (ed.) (Buckingham, Open University Press, 2003)

## A skills progression in ...

| | ... Speaking | ... Listening | ... Group discussion and interaction |
|---|---|---|---|
| **Y1/2** | ❍ Speak clearly and expressively in supportive contexts on a familiar topic.<br>❍ Order talk reasonably and pace well when recounting events or actions.<br>❍ Talk engagingly to listeners with emphasis and varied intonation.<br>❍ Able to use gestures and visual aids to highlight meanings. | ❍ Listen actively following practical consequences, e.g.:<br>– looking at a speaker<br>– asking for repetition if needed.<br>❍ Able to clarify and retain information:<br>– by acting on instructions<br>– by rephrasing in collaboration with others<br>– by asking for more specific information. | ❍ Talk purposefully in pairs and small groups.<br>❍ Contribute ideas in plenary and whole-class discussions.<br>❍ Make and share predictions, take turns, contribute to review of group discussion.<br>❍ Review and comment on effectiveness of group discussions. |
| **Y3/4** | ❍ Sustain speaking to a range of listeners, explaining reasons, or why something interests them.<br>❍ Organize and structure subject matter of their own choice, and pace their talk (including pauses for interaction with listeners) for emphasis and meaning.<br>❍ Adapt talk to the needs of the listeners (such as to visitors or more formal contexts), showing awareness of standard English. | ❍ Sustain listening independently and make notes about what different speakers say, identifying the gist, key ideas and links between them.<br>❍ Able to comment and respond, evaluating a speaker's contribution, or evaluate quality of information provided.<br>❍ Able to concentrate in different contexts, including talk without/by actions and visual aids. | ❍ Sustain different roles in group work (with support from a teacher), including leading and summarizing main reasons for a decision.<br>❍ Talk about language needed to carry out such roles and how they contribute to the overall effectiveness of the work.<br>❍ Reflect constructively on strengths and weaknesses of group talk. |
| **Y5/6** | ❍ Develop ideas in extended turns for a range of purposes.<br>❍ Assimilate information from different sources and contrasting points of view, present ideas in ways appropriate to spoken language.<br>❍ Use features of standard English appropriately in more formal contexts.<br>❍ Make connections and organize thinking. | ❍ Listen actively and selectively for content and tone.<br>❍ Able to distinguish different registers, moving between formal and informal language according to the audience, and emphasize or undercut surface meanings.<br>❍ Able to discern different threads in an argument or the nuances in talk. | ❍ Organize and manage collaborative tasks over time and in different contexts with minimal supervision.<br>❍ Negotiate disagreements and possible solutions, by clarifying the extent of differences, or by putting ideas to the vote.<br>❍ Vary the register and precision of their language and comment on the choices made in more formal contexts. |

Adapted from Primary National Strategy, *Speaking, Listening, Learning: Working with children in Key Stages 1 and 2 Handbook* (Norwich, DfES/HMSO, 2003)

# Collaborative group work

Collaborative group work is an essential part of thinking skills teaching. The opportunity to work with a partner or in a small group is essential. This is where children can explore their own thinking, hear other people's ideas, be tentative, make mistakes, but be supported and encouraged by their peers. This is how an individual develops confidence in new ways of thinking. However, it does not happen automatically. You will need to make time for it, support, nurture and encourage it.

## Plan for it

Thinking about who is going to work with whom, and how, is essential. It won't just happen until the class are used to this way of working, and even then there will be new skills they can develop. Most thinking skills lessons are based on mixed groups that are not based on current levels of attainment. However, you will need to monitor who works well with whom and support the children in working with a wider range of their peers.

## Make it explicit

The children need to know that they are expected to work together, and that you are expecting them to help each other. This needs continual reinforcement with the whole class in the introduction, mini-plenaries and review sections of lessons (praising and reminding groups and individuals helps, too).

## Teach pupils how to work in groups

Not all children find it easy to cooperate. They may well need the first few activities to focus on learning to work together. It is worth making this a part of your learning objectives for speaking and listening (see pages 18 to 19). In one of the early sessions (if you have not done so already), it is worth agreeing class rules for working in groups or a 'working together protocol'. Such an agreement should be phrased positively about what children should do and might include things like:

- Make sure everybody has a turn in speaking
- One person speaks at a time
- Look at the person who is talking (make eye contact)
- Listen actively (positive body language such as nodding or an open posture)
- Speak clearly
- Explain what you mean
- Respond to what other people say
- Make a longer contribution than just one or two words
- Give reasons for what you think
- Make it clear when you disagree that it is with what has been said (with your reasons) and not a person.

However, it is important that the precise wording comes from the children and that the agreement is posted publicly where it will always be visible in the classroom. The children will use it!

## Start small

Pairs are the easiest groups to start with. In Key Stage 1 this should be the main aim. Even very young children should be able to cooperate in pairs, particularly if the cooperation is structured in some way (such as taking turns in a game). Moving from pairs to fours is a good tactic too. A paired task can be reviewed by two pairs to reach agreement, then this larger grouping can form the basis for a further activity.

## Make sure the tasks require cooperation

Consider strategies such as having one recording sheet, or set of resources that need to be shared, or assign specific tasks to each member of the group. As groups get bigger you may need to assign different roles and let the children practise the different skills required (for example, leader, note taker, summarizer, clarifier). In the beginning it is best to use existing friendships as the basis for organizing the groups, but don't let them get too cosy. Learning to work with people who are not close friends is an important skill for life!

> **Suggestions for further reading**
> L. Dawes, N. Mercer and R. Wegerif, *Thinking Together: Activities for teachers and children at Key Stage 2* (Birmingham, Questions Publishing Co., 2000)

# Talking points

## Getting started

How are you going to tackle this?
What information have you got to help you?
What do you need to find out or do?
How are you going to do it? Why that way?
Can you think of any questions you will need to ask?
What do you think the answer will look like?
Can you make a prediction?

## Supporting progress

Can you explain what you have done so far?
What else do you need to do?
Can you think of another way that might have worked?
What do you mean by ...?
What did you notice when ...?
Are you beginning to see a pattern or a rule?

## If someone is stuck ...

Can you say what you have to do in your own words?
Can you talk me through where you are up to?
Is there something that you know already that might help you?
How could you sort things out to help you?
Would a picture help, or a table/sketch/diagram/graph?
Have you talked with your partner/another pair/group about what they are doing?

## Reviewing learning

What have you learned today?
What would you do differently if you were doing this again?
When could you use this approach/idea again?
What are the key points or ideas that you need to remember?
Did it work out the way you expected?
How did you check it?

Remember – one way to ask a question is just to wait!

*Suggestions for further reading*
Association of Teachers of Mathematics, *Primary Questions and Prompts* (Derby, ATM, 2004)

# Thinking skills across the curriculum

There are a number of general teaching strategies that you can explore to support the activities in *Thinking by Numbers*. They are helpful because you can use the same technique in different contexts and develop thinking across the curriculum. Each time you use these strategies you can focus on the children's thinking that you want to develop. The children become familiar with the techniques and can get straight down to the learning involved. The strategies are also useful in assessing the children's understanding. If you first **demonstrate** a technique or approach, you can then set an activity which the children **undertake** to develop their thinking. This is as far as most approaches to thinking skills go. However, if you then set a challenge where the children have to **generate** their own activity based on what they have done, you will see them reveal their understanding of the thinking required. This cycle of **demonstrate**, **undertake** and **generate** ensures that the thinking becomes embedded.

## Odd one out

In this strategy the children are presented with three items and asked to choose one as the 'odd one out' and to give a reason. Items are chosen to ensure that a range of answers is possible. Pupils can also be asked to identify the similar corresponding characteristic of the other two, or features common to all, to develop their vocabulary and understanding. In mathematics this leads naturally on to a discussion of the properties of numbers and to identifying numbers which have a range of properties. It can easily be extended to work on shapes or into other subjects. Selecting three items with different possible reasons is essential. When the children design their own game it is essential that you emphasize that there should be more than one solution or 'answer'. It leads on to identifying common properties that the odd one out lacks.

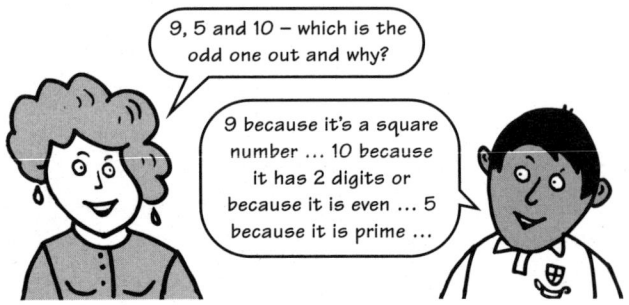

## Living graphs

The strategy involves a graph or a chart as the basis for an activity where the children have to relate short statements to the more abstract structure of a graph. The use of statements that children can understand easily, but which they then have to discuss and interpret, helps them to make sense of both the representation of the graph and the information it is based on. This works well in mathematics and science, but also in other subjects where quantitative information is used, such as history and geography.

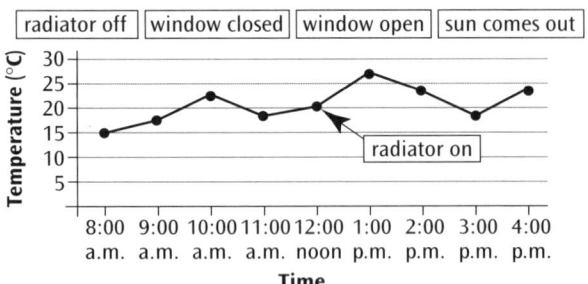

## Sorting strategies

Venn diagrams, Carroll diagrams, grids and matrices are effective strategies to display information visually. They support skills such as collecting, sorting, classifying and organizing ideas and information across the curriculum. Sorting techniques are powerful because they provide examples of what a concept or idea is and importantly, what it is *not*. This approach of using examples and *counter*-examples is a vital teaching tool. Learners tend to over-extend a concept based on their experience, so most children think a trapezium, for example, looks like the roof of a classic drawing of a house, however its defining properties are that it is a quadrilateral with (only) one pair of sides parallel.

## Always, sometimes, never

Another useful strategy is to have a set of statements, such as 'triangles have three sides' or 'multiples of 3 are odd' and ask the children if they are 'always' true, 'sometimes' true or 'never' true. This works well in mathematics and science: in other subjects you may need to set these categories along a continuum to provoke discussion.

| always | sometimes | never |
| --- | --- | --- |

← — — — — — — — — — — — →

As before, asking the children to make up statements that are always, sometimes or never true is a good way to extend the task (and their thinking).

## Reason snap

This can be a good short game to play to encourage simple reasoning language. You display two items (such as numbers or shapes in mathematics) and ask the children to identify a similarity and say, *Snap because …* and to give their reason. This activity can be developed across the curriculum with pictures in geography, artefacts in history, extracts of text in literacy. Thinking can be extended by asking what makes them different or distinctive.

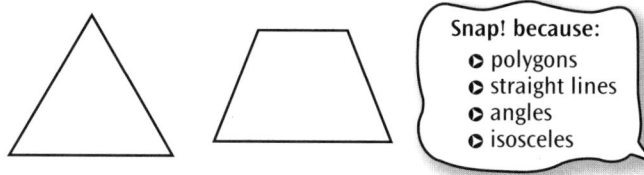

Snap! because:
- ○ polygons
- ○ straight lines
- ○ angles
- ○ isosceles

## Fermi questions

The approach of Enrico Fermi, who was an Italian scientist who used to pose questions to get his team thinking and working together, works well in the classroom. At school a question such as: *How many balloons would it take to fill the school hall?* requires the children to ask a number of related questions along the lies of, *How big is a balloon? How big is the hall?* This particularly develops estimation and approximation skills. Discussion and reasoning is an important part of the process of answering them. Other questions might be: *How many chocolate beans will it take to fill a litre lemonade bottle? What is the total mass of all the children in the school? Or If everyone in school (or the class) lay down in a line from the school gate (or classroom door), head to toe, where would the line end?* Once the children get used to answering questions like this you can ask them to think up their own.

## Banned!

Another strategy uses an approach that involves describing an idea or object without using certain banned words (the Association of Teachers of Mathematics [ATM] have a mathematical version called 'Fourbidden'). This can be used to develop creative use of language to describe familiar ideas and concepts. This strategy works well across the curriculum and can be used to get children thinking creatively about their use of language.

| describe: | without using: |
| --- | --- |
| a square | four<br>sides<br>shape<br>equal |

# Activities in this book

| | |
|---|---|
| **Unit 1**<br>Sort it out!<br>*Information processing skills* | **Little Red Riding Hood (pages 26–29)**<br>The children direct Little Red Riding Hood through the wood to her Grandmother's house and record different routes possible.<br><br>**The Bears' Breakfast (pages 30–33)**<br>After Goldilocks' visit Mrs Bear has to work out if she had enough crockery and cutlery to provide breakfast for the grandparents. The children identify and solve subtractions to help her. |
| **Unit 2**<br>That's because …<br>*Reasoning skills* | **Journey to the Moon (pages 36–39)**<br>Children have to identify calculations to make progress towards the Moon.<br><br>**What's in the ring? (pages 40–43)**<br>The children use a Venn diagram to sort numbers and shapes according to their properties. |
| **Unit 3**<br>Detective work<br>*Enquiry skills* | **How many shapes? (pages 46–49)**<br>The children use their knowledge of shapes to work out how many different shapes they can see in different pictures.<br><br>**Hoopla (pages 50–53)**<br>The children solve a problem based on a fairground show. They work out the possible combinations of three numbers to make a given target. |
| **Unit 4**<br>What if …?<br>*Creative thinking skills* | **The Giant's footprint (pages 56–59)**<br>The children think up solutions to help the Giant work out his height using the size of his footprint.<br><br>**Cinderella (pages 60–63)**<br>The children solve a puzzle to help Cinderella work out how old her wicked stepmother and stepsisters are. |
| **Unit 5**<br>In my opinion …<br>*Evaluation skills* | **Aliens (pages 66–69)**<br>The children work out different ways to pay for a toy and evaluate different methods of keeping track of their working.<br><br>**Funny fish (pages 70–73)**<br>The children investigate different combinations of fish to produce different totals of spots. |
| **Unit 6**<br>Think on!<br>*Using and applying thinking skills* | **Treasure hunt (pages 76–79)**<br>Three pirates hide their treasure in three chests. The children investigate ways to share the treasure and identify criteria for a good solution.<br><br>**Beautiful butterflies (pages 80–83)**<br>The children have to work out how many different butterflies they can make following simple rules. |
| | **The growing caterpillar (pages 84–87)**<br>The children correct graphs showing what the growing caterpillar has eaten during the week then record their stories using similar graphs. |

# Sort it out!
## *Information processing skills*

> **Information processing** – these skills enable pupils to locate and collect relevant information, to sort, classify, sequence, compare, contrast and analyze part/whole relationships. (QCA 2000)

**Overview**

This unit is about working with mathematical ideas and concepts by gathering information. It is about building understanding by actively working with these concepts and ideas. It is about remembering links and making connections to understand what information is relevant. It is also about working with ideas to develop understanding of their meaning by working with patterns and rules, working with definitions and organizing and representing ideas. It is an essential aspect of mathematical thinking. The activities in this unit are designed to help pupils engage practically with ideas and information so as to build their knowledge and understanding of mathematical concepts.

**Strategies**

Information processing skills can be broken down further into the following kinds of behaviours or activities that pupils can do:

- **Find relevant information**
  Remember, recall, search, recognize, identify
- **Collect relevant information**
  Retrieve, identify, select, gather, choose
- **Sort**
  Group, include, exclude, list, make a collection or set
- **Classify**
  Sort, order, arrange *by kind or type*
- **Sequence**
  Order, arrange *by quantity/size/weight*, put in an array
- **Compare**
  Find similarities (and differences), examine, relate, liken
- **Contrast**
  Find differences/similarities, examine, distinguish
- **Analyze part/whole relationships**
  Relate, consider, sort out, make links *between parts and wholes* (e.g. component/integral object (such as the face of a cube); member/collection; portion/mass; stuff/object; place/area; feature/activity; especially in terms of fractions, ratios and the like).

**Questions**

*Can you think of something that might help? What does this remind you of?*
*Give me an example of a ... Is ... an example? Can you give a counter-example?*
*What would come next? What would come before this?*
*Why is it the same/different? What makes it a ...? What is it like? What makes a ... different from a ...?*

# Little Red Riding Hood

BRIEF

In 'Little Red Riding Hood' the children identify possible routes through the wood for Little Red Riding Hood to get to her Grandmother's house. This involves the children recognizing different routes to solve the problem and applying their mathematical knowledge to record their solutions. The game format helps all the children participate and gives you the opportunity to hear their thinking as they talk about their ideas. One of the purposes of the task is to develop flexibility in thinking by incorporating combinations where there is more than one solution. Another is to provide feedback on how confident children are in using particular mathematical language that is new to them (such as 'left' or 'right').

## Key maths links

- Solve mathematical problems or puzzles
- Recognize turns to the left or to the right
- Give instructions for moving along a route

## Thinking skills

- Sorting
- Find relevant information
- Comparing
- Contrasting

## Language

left, right, forwards, backwards, top, bottom, side, middle, over, under, underneath, above, beside, next to, below, right, left, in front, behind, opposite, between, journey, because, same, different

## Resources

PCM 1 (one per pair)
PCM 2 (one per pair)
copy of *Little Red Riding Hood*
colour pencils
counters (optional)

 **Setting the scene**

Remind the children of the story of Little Red Riding Hood. Get them to imagine what it would feel like to walk through a similar wood. Use a technique like 'Freeze framing' to get them to experience how it feels. *How do you know which way to go?* Hold up direction cards (from PCM 2) and ask the children to stand and follow the simple directions: *Right, Left, Forwards.* Have a look at a simple maze and get the children to direct you from the start to the end. *Are there any other ways we could go? How might we record these?* (through different colours)

 **Getting started**

Use PCM 1. In pairs ask the children to see if they can get Red Riding Hood out of the forest. *Remember she can only go forwards, right and left. Can you find different ways out of the forest? Can you find one way? Can you find five other ways?*

### Simplify

*Can you find a way out?* Cut out copies of Red Riding Hood from PCM 1 to allow the children to move her physically through the wood. Alternatively, provide the children with counters to represent Red Riding Hood. Remind the children that she cannot go backwards. Use PCM 1 to draw routes to help them. Draw one route together and talk through directions together: *Left, forwards, right, forwards ...*

### Challenge

Ask the children to think about how they could record the routes on PCM 2. They might want to record the steps so they can compare them easily, e.g. R, L, R, L. *How many turns do your routes have? Are they all the same? Where could the wolf wait so that Red Riding Hood cannot make it safely through the woods?* The children could use counters or cut-outs from PCM 1 to represent Red Riding Hood and the wolf. *How many more routes do you think there will be if Red Riding Hood can also go backwards?*

**Checkpoints**

 ### Watch out for ...

Ensure the children have checked for repeats. Check that they are being systematic in their approaches, e.g. by recording routes with letters (L, R, F, etc) or with different colours, and recognize when all possibilities have been found.

 ### Ask ...

- ❍ *How many different routes did you find?*
- ❍ *Have you tried drawing on the sheet to help you?*
- ❍ *Can you check for repeats?*
- ❍ *Is that the same as ...?*

 ### Listen for ...

Observe the children as they talk to each other. Listen for children using creative approaches to solving the problem and a way of recording them. Are they clear when using the language of position, direction and movement?

 **Moving on ...**

Review some of the routes that the children have chosen. Did they find all possibilities? What was the maximum number of routes? Can they design their own mazes and give instructions to get out? You may also wish to review how well the children worked together. What helped them to work with their partner?

### Where next?

- ❍ Use direct language to move shapes or toys.
- ❍ Talk about things that turn.
- ❍ Make whole turns and half turns.
- ❍ Practise systematic recording during other investigations.

What worked well? Did you intervene only when really necessary?
How did the children work as pairs? Would you prefer the children to
work as a group? Do you need to do more work on these objectives? Were
there any surprises?

**DEBRIEF**

# Red Riding Hood into the maze

**Name** _____ **Date** _____

**Take Red Riding Hood through the maze.**

Finish

Start

**Name** _____  **Date** _____

Write your directions in the table.

| Route | Directions |
|-------|------------|
| 1 | |
| 2 | |
| 3 | |
| 4 | |
| 5 | |
| 6 | |
| 7 | |

## Right

## Left    Forwards

# The Bears' breakfast

In 'The Bears' breakfast' the children help Mummy Bear set the table for breakfast by counting out how many bowls, cups and spoons they will need. Grandma and Grandpa Bear are joining them for breakfast. Unfortunately, since Goldilocks visited, lots of things have been broken. The children are asked to find out how many more of each object they will need. They can then complete a shopping list for Mummy Bear.

## Key maths links

- Place value and ordering
- Understanding addition and subtraction
- Problems involving 'real life'
- Mental calculation strategies

## Thinking skills

- Sorting
- Ordering
- Comparing

## Language

equals, more, less, addition, subtraction, take away, difference, largest, smallest, explain, agree, disagree
How many?
How many more?

## Resources

**PCM 3** (one per pair)
**PCM 4** (one per pair and one enlarged)
**cups, knives, spoons** and **bowls**
**copy** of *Goldilocks and the Three Bears*
**counters** (20 per pupil)

## 1 Setting the scene

Remind the children of the story of Goldilocks and the three bears. Explain that lots of objects have been broken since Goldilocks visited. Grandma and Grandpa Bear are visiting. Show the picture of the empty table (on PCM 4). Explain that Mummy Bear needs to lay the table: *How many bowls will she need? How many spoons? How many knives? How many cups?*

## 2 Getting started

Tell the children that they only have 2 bowls, 1 spoon, 3 cups and 4 knives left. (You can vary the number of items as often as required, to offer different targets for the children.) Give them the target board on PCM 3 and ask them to work in pairs.

### Simplify

*How many knives are there? Have we got enough? What is the difference between 4 and 5? How many more knives do we need?* Match up the knives: 4 + 1 = 5, 5 − 4 = 1, 4 + ? = 5. Repeat the process for bowls and spoons and use counters instead of objects if it helps. Demonstrate how to use the target board on PCM 3 to help the children. *Can you use the target board to show your workings? Can you help complete a shopping list for Mummy Bear?* They can use PCM 4 to record their findings.

### Challenge

Extend the task by setting further challenges related to the story, such as planning a shopping list. They can use PCM 4 to help them. Develop the number investigation by saying that the seven dwarves are going to stop in on their way to the mine: *How many items will they need to get?* (this practises totals to 10) Encourage the children to record their findings.

## 3 Checkpoints

Help the children to say the number sentence using the words 'the difference between ...'. When they are confident ask them to record the subtraction sentences in their books. They can use the target board on PCM 3 to help them. (In this instance, 5 (bears around the table) is the target number with the starting number at the outside. The difference goes in the circle in between.) Demonstrate using the target board with the children and help them to translate the results into number sentences.

### Watch out for ...

Highlight ways that the children use to record systematically and applaud evidence of confident use of subtraction sentences to support their thinking. Children should gain confidence and be able to work more quickly.

### Ask ...

- *How are you recording your work? Which number comes before? After?*
- *What is your target number? How many more? How many more to make ...? What is the difference between?*

### Listen for ...

Listen for the appropriate use of vocabulary, e.g. difference. *I think this because ... I know that because ...*

 **Moving on ...**

Review the solutions with the class. Did they find them all? *Did anyone find a really quick way of working it out? Can you show us?* Change the total on the target board and ask the children to work in pairs to complete the gaps. *Is this a good way to record our work? Why? Why not?*

### Where next?

- Complete the same type of initial activity but involving money. *Bowls cost 5p. How much will I spend? If I have a 20p coin what change will I get? What combination of coins could this be?*
- Link to weights and measures: *One packet of porridge serves three bears. How many packets would they need for five, for twelve? One bottle of milk fills two glasses. How many bottles for five, for twelve?*

What worked well? Was the activity challenging enough? Did you provide enough demonstration before the task started? Did any of the children surprise you? Were they using number lines, counters etc. to support their thinking?

DEBRIEF

**Name** _____ **Date** _____

## Use counters to help you reach the target.

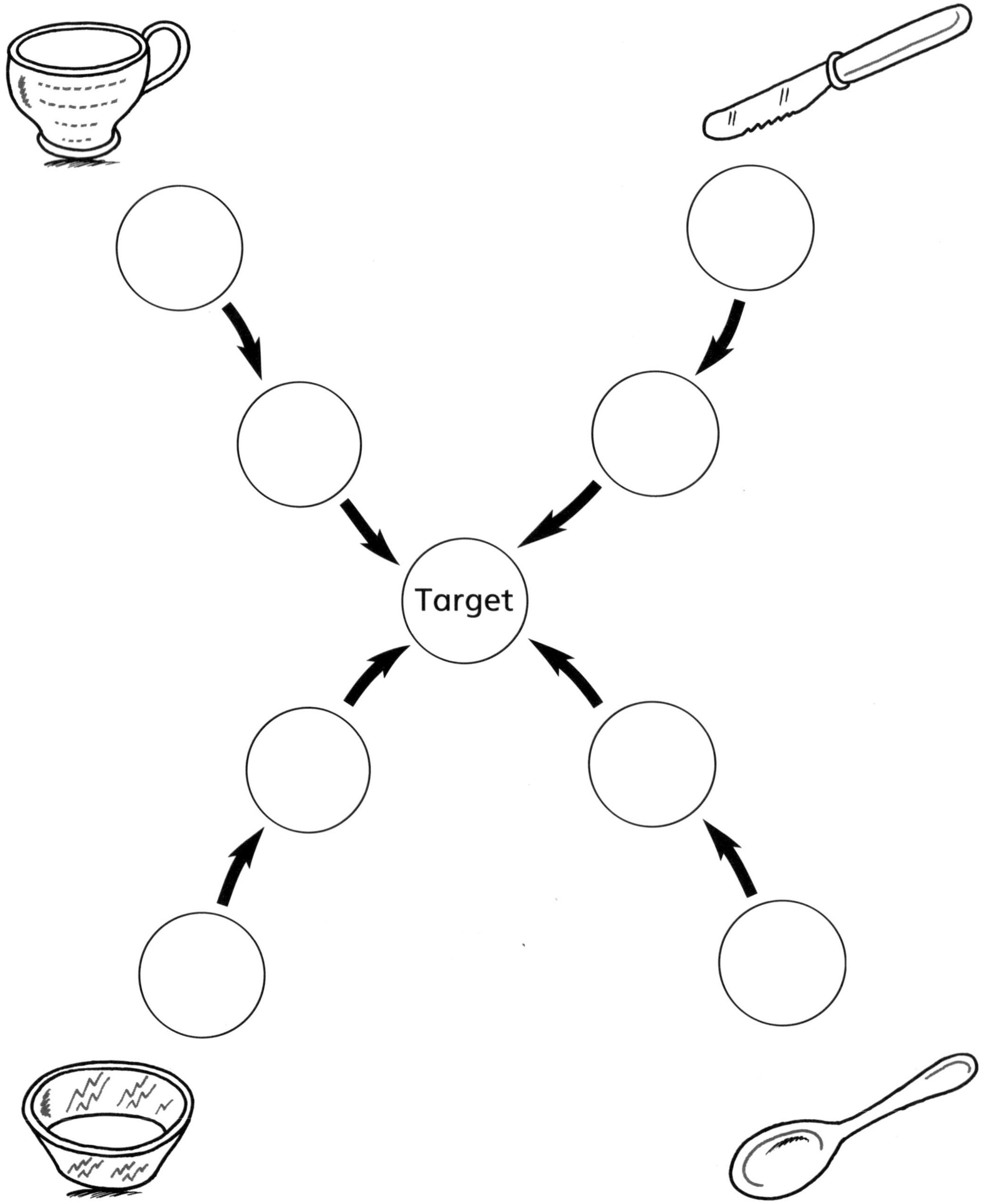

Target

**Name** _____ **Date** _____

| |  | | |
|---|---|---|---|
| | | | |
| | | | |
| | | | |
| | | | |

## Assessing progress

You know that children are developing their skills in information processing when they start to make connections with different mathematical ideas. They should start to show and use this understanding in other lessons. This might be by applying mathematical knowledge in a new situation or it might be in the way that they go about a subsequent task. As their skills in using information develop they should become more precise in the way that they use mathematical language and more systematic in their approach to working and to recording. The techniques that they have used should be developed in other subjects so that their understanding of information processing skills can be transferred to other areas of the curriculum.

## Cross-curricular thinking

### Literacy

A strategy like 'odd one out' can be used to compare characters from fictional genres, such as different heroines from traditional tales.

### Art

The 'odd one out' strategy is also useful for comparing the work of famous artists or to look at similarities and differences in the visual and tactile qualities of materials.

### Science

Venn diagrams are powerful tools in the teaching of classification. This is particularly valuable in the strands of both variation and classification of living things, and materials and their properties.

Little Red Riding Hood, Snow White and Cinderella – who is the odd one out? Why? What makes the other two the same?

### History

Venn diagrams are useful to teach how to compare and contrast in history. Two intersecting sets can be used as a planning tool to identify similarities and differences between different historical periods. Common features go in the intersection and contrasting information on each side.

### Geography

The 'odd one out' strategy and Venn diagrams are both helpful in geography. The former can be used to encourage children to use geographical vocabulary as they talk about what makes three different landscapes or features of the environment similar or different. The latter can be used for sorting pictures of buildings or vocabulary related to the character of places. This way, the children will develop an understanding of these concepts by having examples and counter-examples to talk about in a meaningful context.

# That's because ...

## Reasoning skills

> **Reasoning** – these skills enable pupils to give reasons for opinions and actions, to draw inferences and make deductions, to use precise language to explain what they think and to make judgements and decisions informed by reasons or evidence. (QCA 2000)

## Overview

This unit is about reasoning and logical thinking. Reasoning is an essential aspect of mathematics and underpins the development of theorems and proofs through the use of precise definitions and axioms. For pupils of primary age it is important that they have the opportunity to apply their knowledge and understanding of mathematical ideas and concepts logically and systematically as this will enable them to make connections between different concepts and between different areas of mathematics. This will deepen their understanding and develop their confidence as well as helping them see how mathematics can be used as a practical tool in their daily lives.

Developing reasoning skills is also about developing habits of thinking or dispositions as much as it is about specific logical skills. Of course, just because you are good at reasoning does not mean that you are going to be reasonable. Part of thinking reasonably is also dependent upon your knowledge of yourself and the situation in which you find yourself. This metacognitive dimension is essential if you are going to help your pupils become effective thinkers and not just logical.

## Strategies

Reasoning skills can be broken down further into the following kinds of behaviours or activities that pupils can do:

- **Give reasons for opinions and actions**
  explain, say because, say why
- **Draw inferences and make deductions**
  see links, make connections, infer, deduce, use words like 'so', 'then', 'must be', 'has to be'
- **Use precise language to explain what they think**
  exemplify, describe, define, characterize
- **Make judgements and decisions informed by reasons or evidence**
  form an opinion, determine, conclude, summarize, *especially where there is more than one course of action or possible solution*

## Questions

*Explain why ...? Can you give a reason ...? Because ... So ...*
*Why is ... an example? Is that always/sometimes/never true? What else must be true if ... ? Does it have to be like that? Can you define that? What do they all have in common?*
*What else is like that? What makes you say that? How can you be sure that ...?*

# Journey to the Moon

**BRIEF**

In 'Journey to the Moon' the children solve the problem of getting to the Moon. It's a game that involves exploring sequences of numbers, finding combinations and developing systematic approaches to recording for the different moves. While the game format helps all the children participate, it gives you the opportunity to hear their thinking as they justify their ideas. The task allows you the opportunity to observe and develop their reasoning skills through questioning.

## Key maths links

- Counting, properties of number and number sequences
- Place value and ordering
- Checking results of calculations

## Thinking skills

- Giving reasons
- Working systematically
- Describing

## Language

count on, adding, altogether, one more, two more, total, more, equals What could I try next? I think …

## Resources

**PCM 5** (one per pupil and one enlarged)
**PCM 6** (one per pupil)
**copy** of *Man on the Moon: A day in the life of Bob* by Simon Bartram or non-fiction books on the Moon and travelling through space
**1–6 dice** (one per pair)
**counters**

 **Setting the scene**

The children play a version of 'Snakes and ladders' where they throw the dice to progress but have to write down the addition calculation correctly before they can move.

 **Getting started**

Use an enlarged copy of PCM 5 and demonstrate how to play the game. *Are you all clear how to play? What happens when you land on the end of the rocket jet?* (move up to the top of the rocket) *What happens when you land on the star?* (move down to the bottom of the star) *Where is the first rocket boost? If I am on square 3 and I want to get to 6 what number would I want to roll?* Take responses and agree that 3 more steps would be needed. Encourage the children to play the game a few times. In pairs, using PCM 5 and a dice, ask the children which pairs of numbers would get them from 3 to 9 in two moves. (model 1 + 5 = 6, 2 + 4 = 6 etc.)

### Simplify

In pairs, using PCM 6 and a dice ask them which pairs of numbers would get them from the start to 6 in 2 moves. (model 3 + 3 = 6) *What is the biggest number you could use? What is the smallest number?* Continue until all the numbers are recorded. Help the children to record their moves as a number sentence. *Do the number sentences help?*

### Challenge

*Your counter is on 8. You roll a dice and after two moves you land on 15. Can you find all the ways to get to 15?* The children will need to work in pairs. They can use PCM 6 to aid their recordings. They can use arrows on the number ladders to make number sentences. The number ladders can be used for different starting points.

##  Checkpoints

Are the children recording their progress as number sentences? Look for examples that show methodical working. This is a good opportunity to model different ways of working and invite children who are working systematically to share their methods. How well have they engaged in the task? Was the pairing successful?

How well are they counting on? This is a good opportunity to model the language of adding two scores together.

### Watch out for ...

Some children may count the square that they are on rather than starting with the next square. Model the method of moving on. Prompt the children to check for repeats, e.g. $2 + 4 = 6$ is the same as $4 + 2 = 6$, despite the order of the dice throw. Check that they know when all possibilities have been found. Encourage the use of systematic approaches. Are they clear using the language of addition? Can they record their thinking clearly?

### Ask ...

- *Can you give a reason?*
- *What if ...?*
- *What could I try next?*
- *How do you know?*

### Listen for ...

Observe the children as they talk to each other. Listen for children using appropriate vocabulary and ways of recording addition, e.g. number lines.

## Moving on ...

Investigate pairs of numbers to find all of the possible combinations for different totals. Write a class list in a systematic order to model reasoning. Discuss if there were gaps or repeats. Then, complete the list filling the gaps and deleting the repeats.

### Where next?

- Play the game fully and then ask the children to design their own versions.
- Investigate totals when throwing dice. *Which numbers come up most often?*
- Use a table to show addition totals.

**What worked well? How did the children work as pairs? Were they able to take turns when playing the game? Do you need to do more work on the mathematical objectives? Were there any surprises?**

DEBRIEF

# Rocket to the Moon

## Use counters and a dice to play the game.

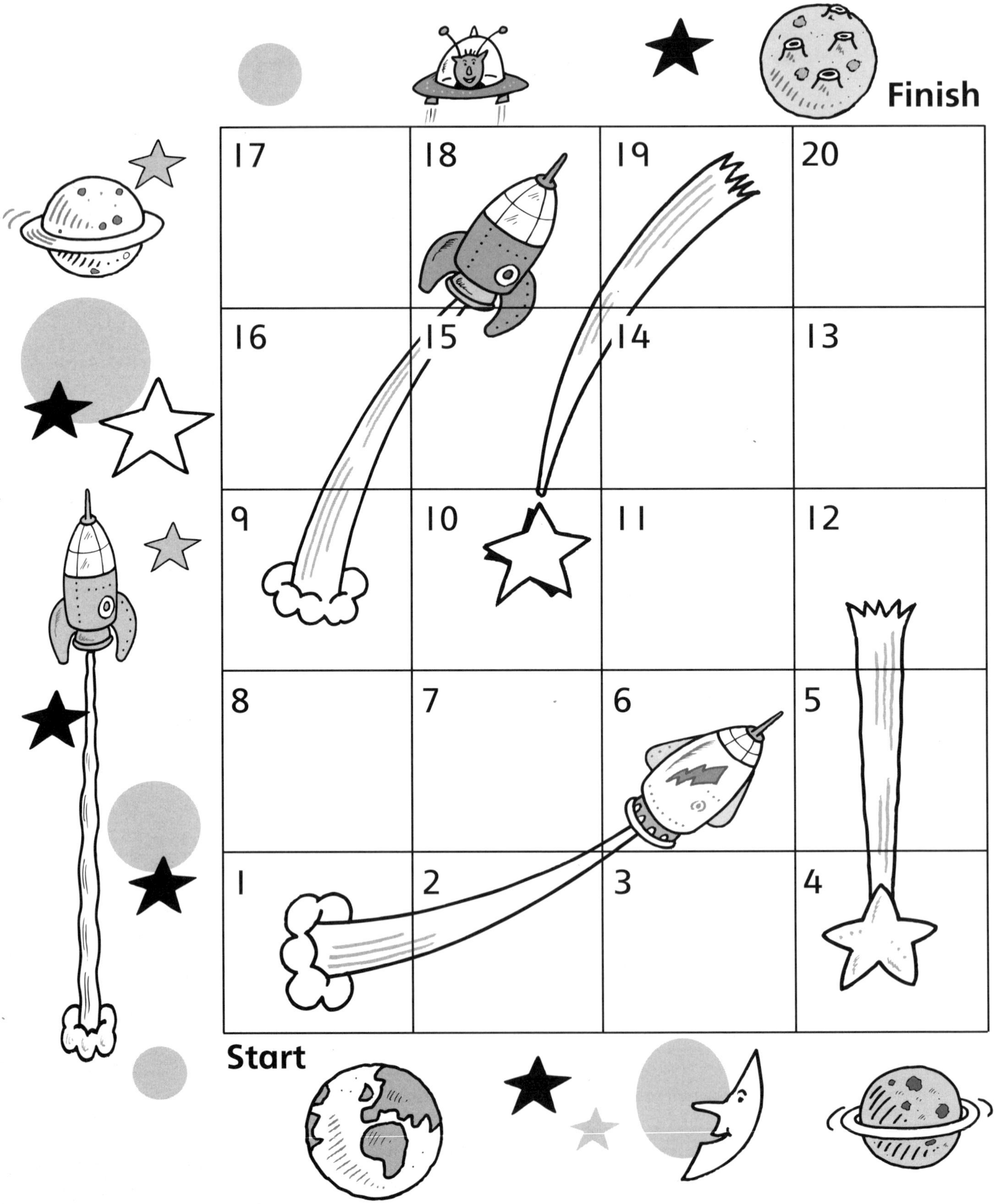

Finish

| 17 | 18 | 19 | 20 |
| 16 | 15 | 14 | 13 |
| 9 | 10 | 11 | 12 |
| 8 | 7 | 6 | 5 |
| 1 | 2 | 3 | 4 |

Start

**Name** _____  **Date** _____

My number sentences to 20.

| 1 |
|---|
| 2 |
| 3 |
| 4 |
| 5 |
| 6 |
| 7 |
| 8 |
| 9 |
| 10 |
| 11 |
| 12 |
| 13 |
| 14 |
| 15 |
| 16 |
| 17 |
| 18 |
| 19 |
| 20 |

| 1 |
|---|
| 2 |
| 3 |
| 4 |
| 5 |
| 6 |
| 7 |
| 8 |
| 9 |
| 10 |
| 11 |
| 12 |
| 13 |
| 14 |
| 15 |
| 16 |
| 17 |
| 18 |
| 19 |
| 20 |

**Workings**

# What's in the ring?

**BRIEF**

In 'What's in the ring?' the children use a Venn diagram to sort numbers and shapes according to their properties. They then hide the labels to make a puzzle for their partner to solve. This develops their recognition of similarities in the properties of numbers and shapes and develops their reasoning skills through their explanations of the task.

## Key maths links

- Reasoning about numbers and shapes
- Counting properties of numbers and number sequences
- Sort shapes and describe some of their features
- Organizing and using data

## Thinking skills

- Recognizing properties of numbers and shapes
- Classifying numbers and shapes
- Identifying similarities

## Language

1-digit/2-digit, always, different, double, even, odd, property, same, teens number

## Resources

**PCM 7** (one per pair)
**PCM 8** (one between two and enlarged if using large number cards)
**0–20 number cards**
**shapes** (regular and irregular)
**scissors**

## 1 Setting the scene

Use an enlarged version of PCM 8 to sort numbers into the ring. Make sure some numbers are sorted but left outside the ring (e.g. odd numbers inside and even numbers outside). Ask the children for other numbers that could go inside or outside the set. Repeat this until you have used most of the different labels. Next, ask the children to close their eyes while you make a puzzle for them to work out. Hide the label so the children cannot read it and place some numbers inside and outside the set. Ask them to open their eyes and see if they can work out what the label should be. Encourage them to explain their reasoning.

## 2 Getting started

Tell the children to work with a partner and to choose a label from PCM 7 for the set. Ask them to take some number cards each and take it in turns sort them using the circle. Get them to explain to their partner what they are doing. They could record what they do on a reduced version of PCM 8 (with four rings per sheet). Once they are confident with using the labels ask them to make a puzzle for each other by doing a sort, then turning the label over. Get them to ask their partner to guess the label.

### Simplify

Choose the labels and collections of numbers for children to use. Try familiar properties like odd, even and more than 10 or less than 10. Using collections of shapes with labels like 'triangle' and 'square' is also less challenging.

### Challenge

Only place a few numbers inside the circle (e.g. 2, 4, 6). Ask the children what labels could be used (even, less than 10, doubles, 1-digit). Add a number outside the ring and ask what labels could now be used and why. Ask them to make up similar puzzles.

## 3 Checkpoints

Ask them to talk to their partner about what they are doing and explain their reasons when they position the numbers inside or outside the ring. Encourage them to place numbers outside as they will find this more difficult.

### Watch out for ...

Children may have difficulty in seeing when they do not have enough numbers sorted to be certain of the label. Negative criteria are harder than positive (i.e. it is easier to put things in the set using the label than spotting numbers that go outside).

### Ask ...

- ❍ *Is it even or not even?*
- ❍ *Does it go inside or outside?*
- ❍ *Why have you placed that number there?*
- ❍ *Can you think of another number that could go there?*

### Listen for ...

Choose some of the pairs to model the mathematical language you are looking for to the whole class. Look for evidence of the children being systematic in the way that they are working together and sorting the numbers.

## 4 Moving on ...

See if they can explain how they solved the puzzles and if they had any strategies to do so (such as using the numbers outside to check). Ask some of the children to try out their puzzles on the rest of the class while they try to use the strategies you have just discussed.

### Where next?

- ❍ Extend the challenge by asking the children to identify other numbers to go inside and outside the set (up to 100).
- ❍ Use the materials to sort shapes in mathematics.
- ❍ Try a similar sorting task in science (such as living things or materials and their properties).

What worked well in this activity? Were the children able to sort effectively? Could they all work out the labels or did some children find this more difficult? How did they cope with puzzles? Did anything surprise you?

DEBRIEF

| | |
|---|---|
| odd | even |
| 1-digit | 2-digit |
| less than 10 | more than 10 |
| more than 20 | teens number |
| doubles | curved lines |
| straight lines | square |
| circle | star |
| triangle | three sides |
| four sides | |
| | |

**Name** _____ **Date** _____

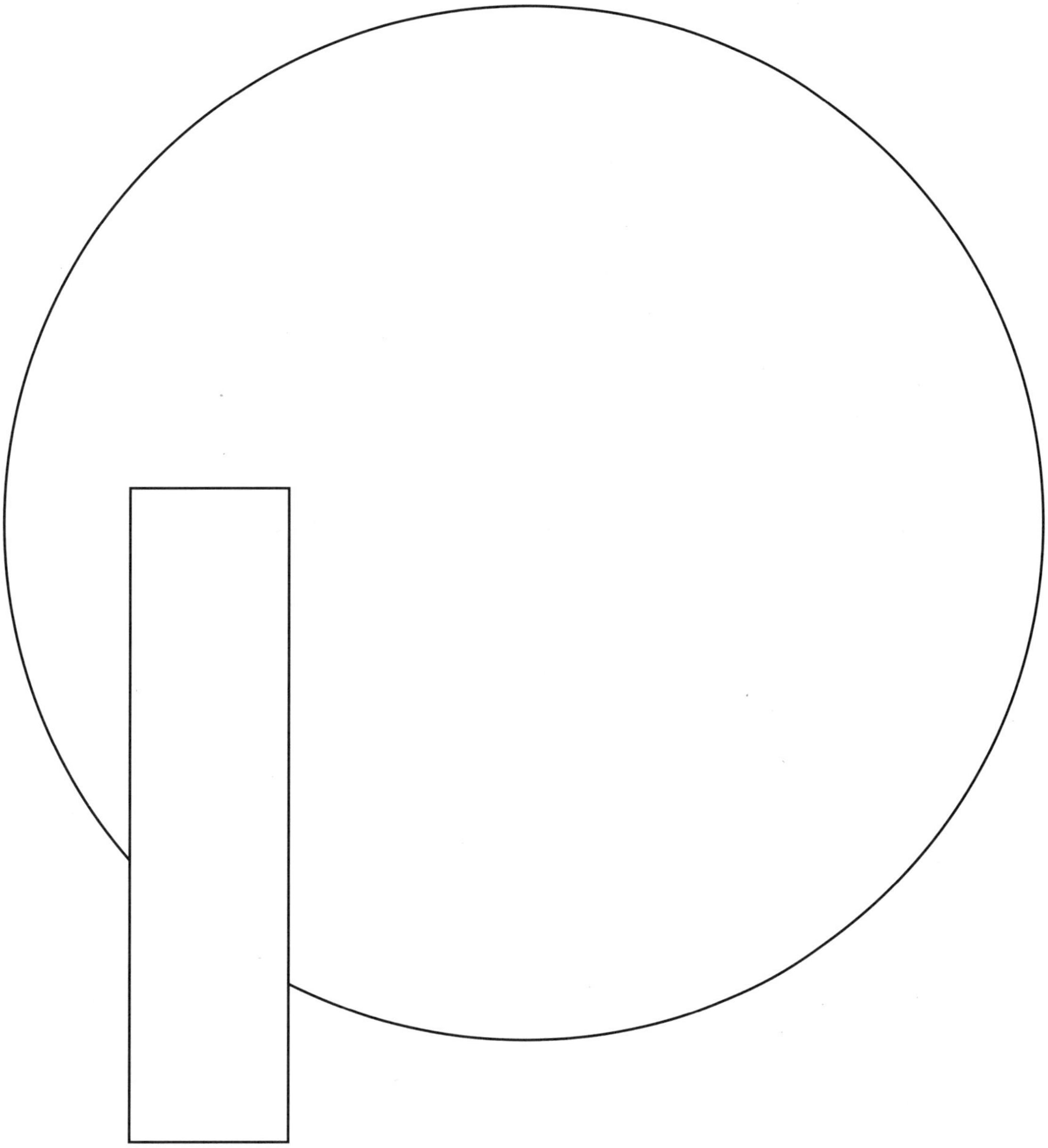

## Assessing progress

You know that children are developing their reasoning skills when they start using words like 'because', 'then' and 'so' in their discussions and their responses to your questions. They may also start to ask each other 'why?' questions and seek explanations from each other (and from you). Giving reasons as part of explanations then becomes a routine part of thinking lessons. Once you start to ask children why (or to ask another child why a response either was or was not correct) you will be able to assess the reasons in their responses. You need to ensure you ask children to justify correct and incorrect responses otherwise they will 'read' your question as meaning they have made a mistake if you only ask 'why?' when an answer is wrong. Once children get used to this you can simply wait encouragingly or say 'because ...?' to get them to extend their replies to your questions to assess their reasoning skills.

## Cross-curricular thinking

### Science

Asking a question such as, *What will happen if?* is a good starting point for scientific reasoning. *What will happen if you put a tea cosy over an icy drink? Will it warm up faster or more slowly? What will happen if you drop a football and a cannonball from the top of a tall building? Will the cannonball reach the ground first?* Using thought provoking questions like these can stimulate scientific reasoning (as well as revealing children's thinking about scientific concepts).

### History

A strategy like 'odd one out' can also be used to develop reasoning skills as the children are asked to give reasons for their choice of an 'odd one out' and can be encouraged to distinguish between historical and non-historical reasons. Choose three famous people and ask children to identify an odd one out with a historical reason.

### Literacy

Justifying choices of words and phrases is a good way both to develop reasoning, and model thinking about composition. Asking a series of questions such as, *Why did you choose that adjective or powerful verb? What others did you consider? Why did you reject those?*, not only gives children the opportunity to give their reasons, but to make them explicit for others to hear.

### Geography

Geographical enquiry is supported by reasoning as children express they views about places or changes to the environment. They can use a technique, such as identifying 'Plus, minus and interesting' points to compile a table, then justify the points they have identified with reasons.

# Detective work
## *Enquiry skills*

> **Enquiry** – these skills enable pupils to ask relevant questions, to pose and define problems, to plan what to do and how to carry out research, to predict outcomes and anticipate responses, to test conclusions and improve ideas. (QCA 2000)

**Overview**

Enquiry skills are as much a way of working or developing particular habits of mind which keep a range of possibilities open for as long as possible. The process of enquiry is about being flexible, looking for alternatives and testing a range of possible solutions. In mathematics these are essential skills as enquiry develops an understanding of relationships and connections that may not be immediately obvious.

The process of enquiry is at the heart of learning. It is only when you can identify what you need to know, go through a process of finding out and be able to recognize when you have found a solution that you can undertake independent learning. Enquiry skills can, therefore, best be developed in situations where it is not possible to see a solution from the outset and where children will benefit from working together.

There are good opportunities for speaking and listening in presenting the results of an investigation or enquiry. Enquiry lessons are also excellent for review and reflection about the process of learning.

The challenge for the teacher is at the beginning and end of the enquiry process. It is difficult to instruct children in how to ask relevant questions without directing them to a particular investigation or mathematical problem. Similarly, it is difficult enough for pupils to recognize that they have come up with a solution to an investigation, without them realizing that it is a good solution. Identifying what would be a better answer is even more difficult - challenging even for adults! Enquiry skills are also therefore, about developing more systematic habits of questioning as well as the specific skills in solving a problem.

**Strategies**

Enquiry skills can be broken down further into the following kinds of behaviours or activities that pupils can do:
- **Ask relevant questions**
  Enquire, be curious, ask, probe, investigate
- **Pose and define problems**
  Frame, propose, suggest, put forward an idea
- **Plan what to do and how to research**
  Think out, plan, sketch, formulate or organize ideas
- **Predict outcomes and anticipate responses**
  Suppose, predict, guess, estimate, approximate, foresee
- **Test conclusions and improve ideas**
  Experiment, test, improve, refine, revise, amend, perfect

**Questions**

*Show me how you could ...? What might work? What ideas have you got? What is a good question to ask? How could you find out? How could you check? Any predictions? What is your best guess? What are you expecting? About how much will it be?*

# How many shapes?

**BRIEF**

The 'How many shapes?' activity makes useful links with art and artists. It encourages the children to think about using only one shape to draw with. It requires them to apply their knowledge of shapes and to use their enquiry skills to find out how many shapes they can identify in a piece of art.

## Key maths links

- Reasoning about numbers or shape
- Shape and space

## Thinking skills

- Giving reasons
- Explaining
- Making connections
- Working systematically

## Language

shape, pattern, triangle, rectangle, sides, corners, flat, straight, make, build, draw

I think …

## Resources

**PCM 9** (one per pair)
**PCM 10** (one per pair)
**flat shapes/pre-cut sticky shapes,**
**stamping** or **printing shapes**
**posters** of Cubism by Picasso and Cezanne

 **Setting the scene**

Show the class some posters and postcards of works of art by various artists, especially of the Cubist movement, such as Picasso and Cezanne. Encourage the children to point out any geometric shapes they see. Then show the children an enlarged copy of PCM 9. You will need to explain that Mr T Riangle is a very happy man. *Can anyone tell me why?* (he loves triangles) Tell them that, unfortunately, Mr T Riangle doesn't know how to count and needs help to count all his triangles. The more we find the happier he is. *Can we help him?*

 **Getting started**

Give the children a copy of PCM 9 each and ask them to work in pairs to find a solution. When a pair find some of the body combinations (there are 9 triangles on his body alone) bring the class together and ask for the pair to describe and explain their findings. Take the opportunity to share different ways of keeping track and recording. (there are 17 triangles used to make Mr T Riangle)

Using a collection of flat shapes, pre-cut sticky shapes, stamping or printing shapes ask the children to develop this idea. *Can you now make a vehicle for Mr T Riangle to get to work in? How many triangles did you use? Does your partner agree?*

### Simplify

Suggest to the children that they cut out Mr T Riangle and all the shapes that he is made from so they can move the pieces. (Use a laminated version, if easier.) Encourage them to draw around the triangles to aid their understanding and assist their counting.

### Challenge

Once the children have completed PCM 9 give them PCM 10. Check that they know what a rectangle is. They can then deduce how many rectangles can be seen in the rocket. (there are 18 rectangles) Then ask them to develop this idea. *Can you make your own rocket using rectangles? Can you make an alien using rectangles? Ask your partner to guess how many rectangles you used.*

## 3 Checkpoints

Look for pairs who are working well together and confidently identifying the shapes. Encourage pairs who are recording in a systematic way. Help only when necessary.

### Watch out for ...

Encourage children who can confidently identify and use the shapes in different orientations.

Look for children who have developed original approaches to recording. Look for confident use of vocabulary

### Ask ...

- ○ *Have you found all the triangles?*
- ○ *Have you found all the rectangles?*
- ○ *How did you make ...?*
- ○ *Would it be easier to use other shapes as well?*
- ○ *How is a rectangle different from a triangle?*

### Listen for ...

Some children may voice sensible reasoning: *Two small triangles put together like that make a larger triangle, which makes three triangles in total.*

## 4 Moving on ...

Review the shapes the children have been making. Work together to estimate then count the shapes used. *If you were an artist would you like to use only one shape? Why? Why not?*

### Where next?

- ○ Make pictures and patterns using straws, modelling clay and pipe cleaners, then link this to 3D shapes.
- ○ Look at repeated patterns and identify the shapes used.
- ○ *What can you make with only circles?*
- ○ Use two different shapes and explore the new patterns and designs the children can make.

Were the children able to complete the tasks successfully? Did you challenge them enough? Did you have enough support for the less able? Did any of the children surprise you? Do you need to reinforce any mathematical knowledge in further lessons?

DEBRIEF

# Mr T Riangle

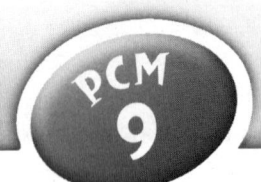

Name ______________________ Date ____________

Can you guess his favourite shape?
How many can you see?

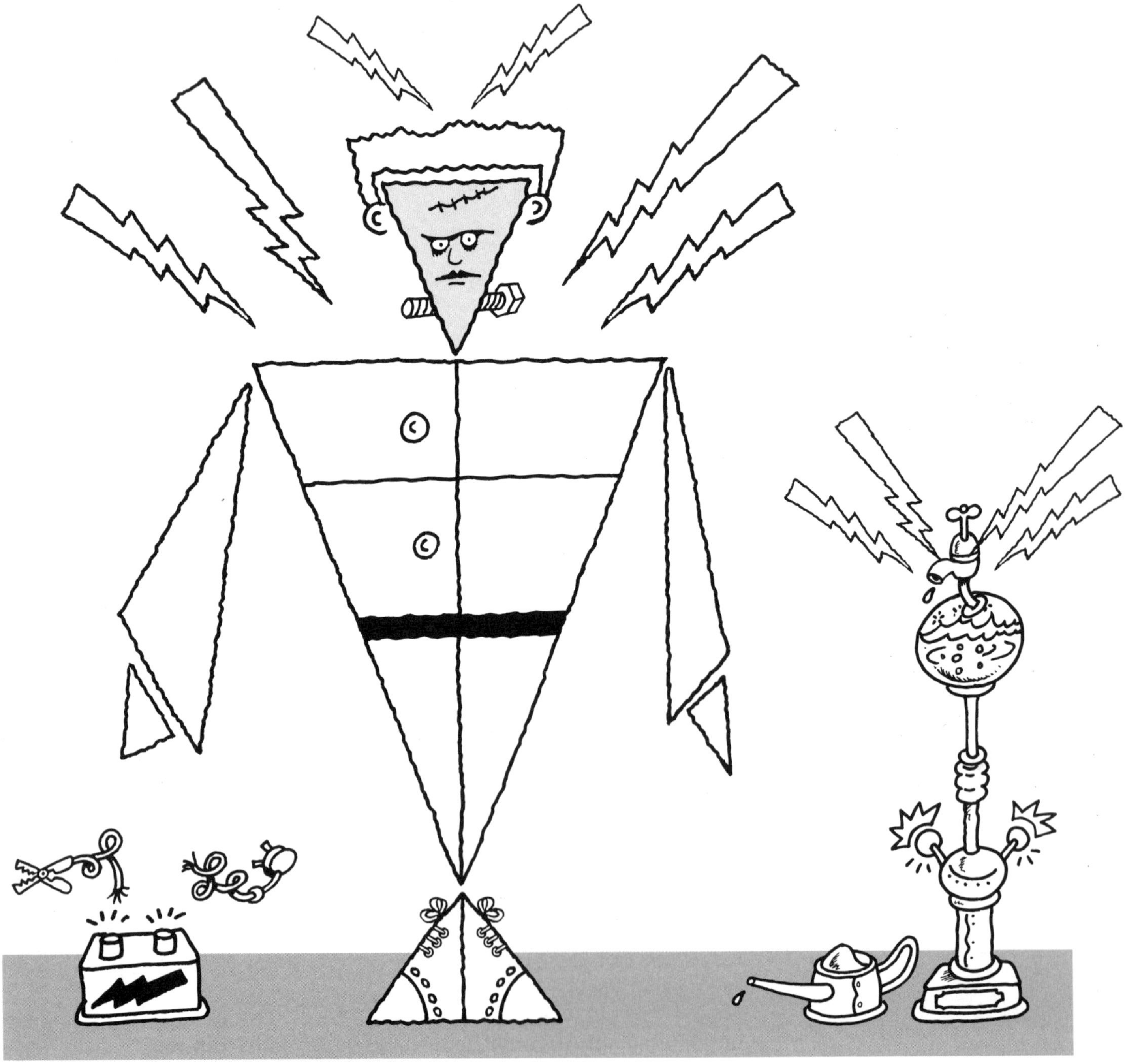

**Name** _____  **Date** _____

How many rectangles?

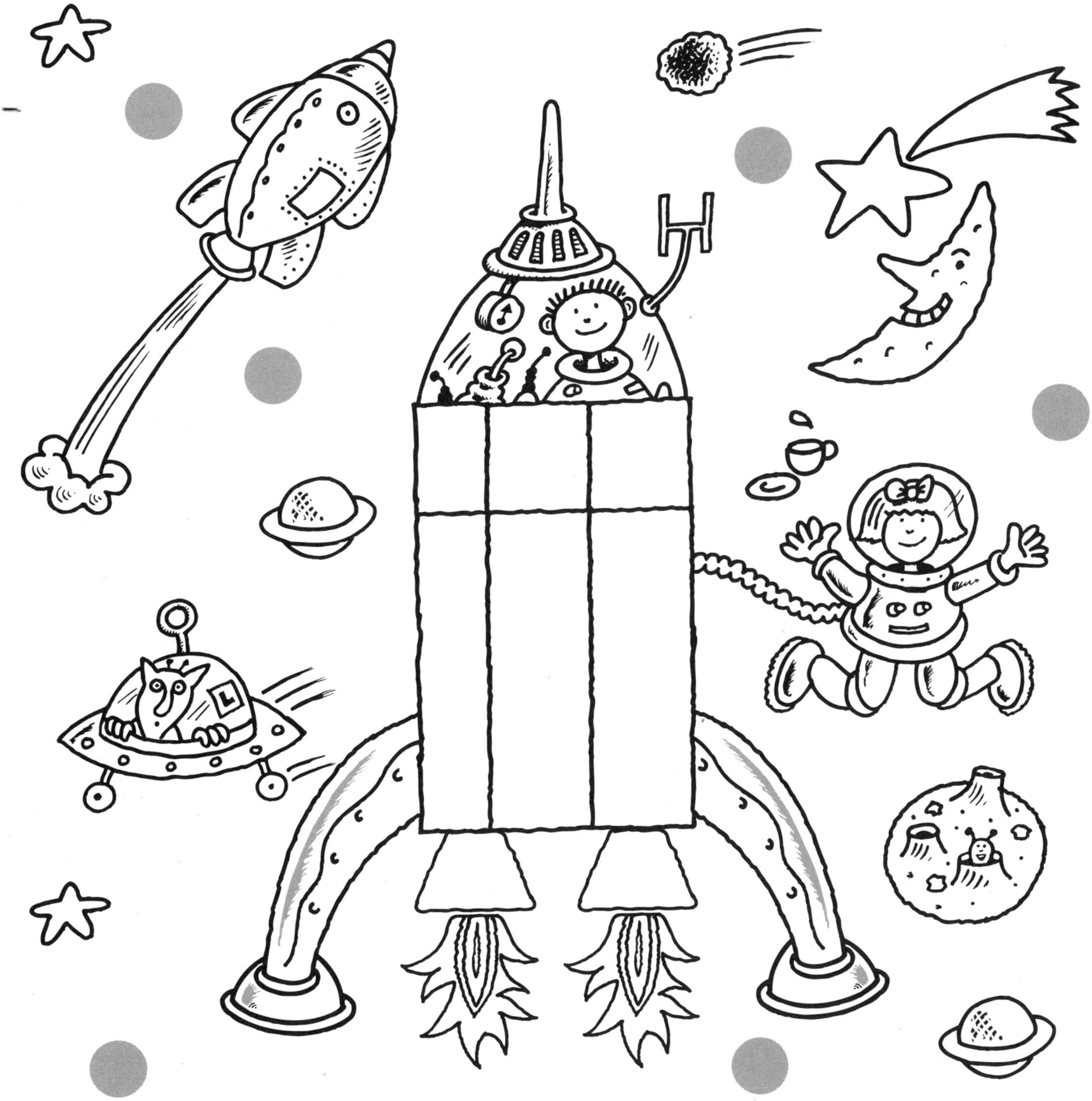

Can you make a rocket using rectangles?
How many rectangles?

# Hoopla

**BRIEF**

In 'Hoopla' the children are presented with Huang who is at the fair and has three hoops to throw. They need to see if they can make various target numbers using three throws. This game allows the children to investigate combinations and to consider how they will record their work. It will also give you the opportunity to listen to their reasoning behind statements such as, *I can make 11 but not 10.*

## Key maths links

- Estimating and rounding
- Rapid recall of addition and subtraction facts
- Making decisions

## Thinking skills

- Investigating
- Problem-solving
- Proposing ideas
- Testing solutions

## Language

count, predict, guess, too many, too few, enough, altogether, equals
How many? How many more to make …?

## Resources

**PCM 11** (one per pair and an enlarged copy)
**PCM 12** (one per pair)
**role-play area** set up as a hoopla stall
**whiteboards**
**hoops**

 **Setting the scene**

Show the children an enlarged copy of PCM 11. Ask them what they think Huang has to do. Explain that he can score either 1, 2, 5 or 10. Use three hoops and four pieces of paper: one labelled 1, the next 2, the next 5 and the other 10. Ask a child to throw a hoop. *Where does it land? What is the total after three hoops? What is the lowest score he could get? What is the highest?* Demonstrate how to record this throw on PCM 11.

 **Getting started**

Ask the children to work in pairs. They will need a copy of PCM 11 and 12 between them. Can they make the target numbers on PCM 12? What other target numbers can they make? You might want to suggest a limit to start, e.g. the numbers between 1 and 10.

### Simplify

Use the hoops and paper until you are clear that the children have understood the concept. You can then work with counters in each of the boxes on PCM 11 to aid their addition. Take the opportunity to help them by starting with the highest number and then counting on. Also, reinforce the benefit of informal jottings/workings. *Can you make all the numbers between 1 and 10?* (not 1, 2 and 10) *Can you make all the numbers between 1 and 20?* (not 1, 2, 10, 18, 19) *What is the lowest number you can make?* (3) Use PCM 11 to support their recording, if appropriate.

### Challenge

*What is the lowest number you can make?* (3) *What is the highest number?* (30) *Can you make all the numbers in between?* You can also introduce the idea that Huang might miss. *How would we record this? Can Huang make all the numbers to 30 with four throws?*

## 3 Checkpoints

Sit with a group and listen to their conversation. It is important to see who is working well by sharing their ideas and listening to others. Through mini-plenaries introduce the idea that there are totals that can't be made (1, 2, 10, 18, 19). Share examples.

### Watch out for ...

This is a good opportunity to model recording using number sentences. Look for examples from the children and get them to share these with the rest of the class. Does anyone need extra resources, such as counters, to aid their recording?

### Ask ...

- *What is the smallest number we can make?*
- *What is the largest?*
- *Can anyone make 10 with three throws? Why not?*
- *You can get your money back if you get 18 with three throws; can anyone get their money back?*

### Listen for ...

Some children will make sensible deductions, e.g. *The lowest score I can make is 3 because 1 + 1 + 1 = 3.* Identify the children who are working well together, and the children who will need more practice in the other mathematics activities to support these skills.

## 4 Moving on ...

Review the numbers that can't be made. *Has anyone been able to make 1 ,2, 10, 18 or 19? Why not?*

### Where next?

- *What would happen if Huang could throw the hoop four times? Can Huang make 10 now? (5 + 2 + 2 + 1) Or 18? Or 19?*
- *If Huang could throw the hoop as many times as he liked, could he make all the numbers from 1 to 30?*
- *If the scores for the hoopla were now 2, 4, 8 and 3 what scores could Huang make with two throws? What scores can I make with three?*

**What worked well? Did you have enough resources to support the less able? Do the children need more help with recording their work and can you develop this in the maths lesson?**

DEBRIEF

**Name** _____  **Date** _____

What score can Huang get with three throws?

| First throw | Second throw | Third throw | Total |
|---|---|---|---|
|  |  |  |  |
|  |  |  |  |
|  |  |  |  |
|  |  |  |  |
|  |  |  |  |
|  |  |  |  |
|  |  |  |  |
|  |  |  |  |

# At the hoopla stall

**Name** _____ **Date** _____

HOOPLA

2    1    5    10

PRIZES

Money back on **26**

Can you make the target score with three throws?

| First | Second | Third | Target Scores |
|---|---|---|---|
| | | | 3 |
| | | | 4 |
| | | | 5 |
| | | | 13 |
| | | | 14 |
| | | | 15 |
| | | | 20 |
| | | | 21 |
| | | | 22 |
| | | | 26 |

**Thinking by Numbers 1 • Unit 3: Detective work • Enquiry skills**

## Assessing progress

Evidence that children are making progress in developing enquiry skills can be gained by observing the way that they are working. Enquiry is as much a habit, or an attitude, of keeping a range of possibilities open for as long as possible. Being flexible, looking for alternatives and testing a range of possible solutions are therefore good indications that enquiry skills are developing.

## Cross-curricular thinking

### Literacy

Another variation on the 'Living graphs' (page 22) strategy in developing the understanding of narratives, both in fiction and non-fiction texts (such as historical narratives), through discussion and enquiry. The graph is replaced with a 'fortune line' about a character's feelings or mood. The children place statements from the narrative on the graph. To do this they need to sequence the text and empathize with the character. Investigating a number of similar narratives (such as traditional tales) will show that they tend to have a similar shaped graph, reflecting the narrative structure and the use of repetition to develop suspense (in *The Three Billy Goats Gruff* and *Little Red Riding Hood*, for example).

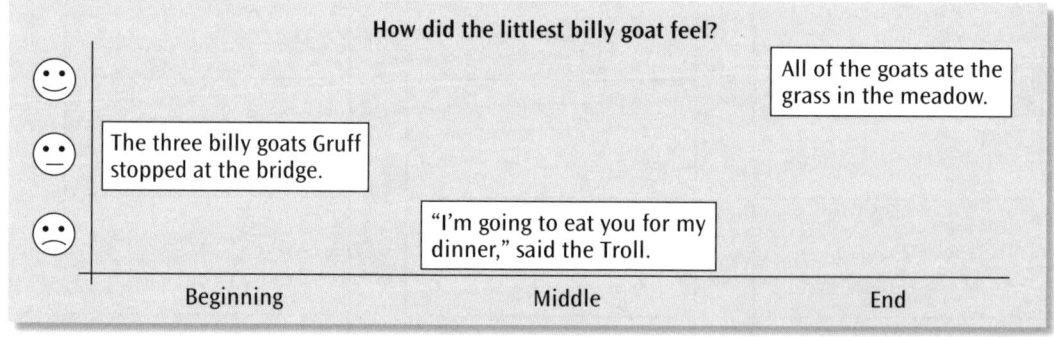

How did the littlest billy goat feel?

- All of the goats ate the grass in the meadow.
- The three billy goats Gruff stopped at the bridge.
- "I'm going to eat you for my dinner," said the Troll.

Beginning        Middle        End

### History

Fortune lines can also be used in historical enquiry particularly to develop empathy, the statements can either come from real historical figures (the diaries of Samuel Pepys and Anne Frank are good sources) or characters created for the task (such as a child miner in Victorian times).

### Science

Developing scientific enquiry means the children must think up questions that can be investigated. An approach called 'Philosophy for children' has been shown to encourage children to develop questioning skills. It uses a stimulus as a starting point, commonly a familiar story, but it can be a poem or a picture, that the children think up questions about. They then select one to answer in a class discussion called a 'community of enquiry'. It is possible to extend this into science where questions can be investigated and you can challenge the children to work out how they could find out the answer.

### Geography

This approach can work in geography, particularly when a photograph of a an interesting landscape is used as the starting point. After a discussion of what the children think, their motivation to find out is likely to be enhanced.

# What if ...?

## Creative thinking skills

> **Creative thinking** – these skills enable pupils to generate and extend ideas, to suggest hypotheses, to apply imagination and to look for alternative, innovative outcomes. (QCA 2000)

### Overview

Creative thinking is the kind of thinking that produces new insights, approaches, or perspectives. It is essential in education that learners see that they can come up with new ideas or suggestions which help their own thinking as well as stimulating the thinking of others. No one expects a 7- or 11-year-old to come up with something unique in the history of human development, but unless we value the creativity that young children naturally have they will stop thinking creatively and rely on reproducing ideas they have been given by others.

Creativity is often *not* associated with mathematics in schools, but thinking up new solutions to problems, seeing new connections, or thinking of more efficient or effective alternatives is what mathematicians do. It is not necessary for the ideas to be completely original, just new for the individual pupil or shared with the class for the first time, or it might be that ideas or concepts are seen in a new or unusual way. It is important that pupils feel comfortable in order to be creative. They need to have confidence that their ideas will be accepted and that there is a range of possible answers or solutions to a problem or issue. The aim is to encourage pupils to think up a range of ideas, to have new thoughts or ideas (at least for them) or to extend and develop other people's ideas.

There are a number of techniques and approaches to support creative thinking such as brainstorming, thinking of analogies, visualizing or picturing possibilities. What all these techniques have in common is an emphasis on the flow of ideas. This means that in the early stages of supporting creative thinking it is essential to be uncritical to ensure that thinking is not too restricted.

### Strategies

Creative thinking skills can be broken down further into the following kinds of behaviours or activities that pupils can do:
- **Generate and extend ideas**
  Brainstorm, think up, develop, extend
- **Suggest hypotheses**
  Suppose, surmise (*use phrases like 'how about ...?', 'it could be ...'*)
- **Apply imagination**
  Design, devise, visualize, elaborate
- **Look for alternative, innovative outcomes**
  Think laterally, fancy, guestimate, invent

### Questions

*Can you imagine? What would that look like? How could you change it to make it a ... ? Can you think of a question you could ask? Go on ... What will the answer look like? Another idea? And another ...*

# The Giant's footprint

In 'The Giant's footprint' the children are asked to find out how many footprints are equal to their height. The investigation enables the children to make predictions, not only about how big their feet are but also about how tall they are. It enables them to estimate who has the biggest feet and by how much, and to compare foot size with height. From this comparison they work out how tall a giant might be from measuring his footprint.

## Key maths links

- Estimating and rounding
- Measures
- Problems involving 'real life'

## Thinking skills

- Think up ideas
- Invent solutions
- Imagine possibilities

## Language

long, short, tall, high, longer, longest, guess, roughly, nearly, close to, too many, too few, ruler, matchsticks, centimetre, metre
What if …? What could I try next? About the same as …

## Resources

**PCM 13** (one per pair)
**PCM 14** (one per pair, enlarged)
**rulers**
**measuring equipment**
**paper** (in a roll)
**scissors**

 **Setting the scene**

Explain to the children that the Giant from *Jack and the Beanstalk* is really stuck. (A giant letter asking for help - written to the class - is a great way of engaging the children. Enclose an enlarged copy of the footprint on PCM 13.) He needs to work out how tall he is so he can get a new passport. He is sure that the class working together can help. *Can you help? Do you think we should find out more about ourselves before we can help the Giant? What do you think we should measure?* Brainstorm the possible ways to investigate the problem. Show the children a copy of PCM 13. *How do we know how big our feet are? How can we measure them? Can you do this on your own?*

 **Getting started**

Give each pair of children a copy of PCM 13. Use the footprint outline to talk about how they can draw around their foot (this is much easier if they keep their shoes on!) and use this as a template. Then they can draw around themselves to obtain a record of their height. Bring the class together. *How can you help each other? How tall are you in footprints?* You will then need to demonstrate how to use the footprint template they have made to measure their height. The children will find out their height is a multiple of their foot(print) length. Take the opportunity to discuss methods for measuring.

### Simplify

Suggest to this group that they use their footprint to draw comparisons, e.g. *Who has the biggest feet in the group? Who has the smallest feet? Can we put them in order, smallest to largest?* If the group is ready they can then use their templates to measure their height in 'number of footprints'. Offer teaching assistant support, where necessary. Suggest that they compare their height in number of footprints with the Giant's.

### Challenge

The children can produce their foot templates on PCM 13 and then work out their height in 'number of footprints'. They can use this information creatively to work out the height of the Giant. Present this group with an enlarged version of PCM 14 to work with. *What are you going to use to measure? How large is it? If you are 7 footprints tall, how many footprints tall is the Giant? Can we work out his height now?*

 **Checkpoints**

Look for examples that show systematic working. Ask these children to share their methods. How well have they engaged in the task? Is paired work successful? How well are they counting on? Can they apply their footprint to height relationship to the Giant?

### Watch out for ...

Some children may get confused with adding, offer them support with measuring equipment and model methods to help them with their calculations.

### Ask ...

- *How are you recording your work?*
- *How tall are you in footsteps?*
- *Did you work well together? Why? Why not?*
- *How can we write that?*
- *What are you going to use to measure?*

### Listen for ...

Observe the children drawing their footprints. Listen to them measuring their height in footprints. Listen for comparisons with the Giant's height: *If my height is seven times my footprint, I bet it's the same for the Giant. I think the Giant is taller than the classroom.*

 **Moving on ...**

Review the solutions found. Did they find out how tall they were in footprints? Ask the 'Challenge' group to give their solution to the rest of the class. Did they find out how long their foot/footprint was? *Can we write back to the Giant and tell him how tall he is?* Write a short reply to the Giant giving him his height in the units that the children used.

### Where next?

- The children can model the Giant's height by making a row of footprints.
- Display a comparison of foot size in the class by glueing the children's templates onto a graph to produce a pictogram.
- Explore other graphs to show more information about the class and look at answering questions from the information, e.g. *How many children have blue eyes? How many more have brown eyes?*
- Introduce the children to the idea of standard units for measuring. *Can we compare footprints?*

**What worked well? Was this activity challenging enough? Do you need to cover more work on measures? Are there any possibilities to extend this activity further?**

DEBRIEF

# My footprint

**Name** _____ **Date** _____

## How big is your foot? How tall are you in footprints?

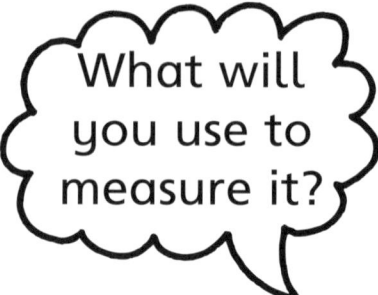

What will you use to measure it?

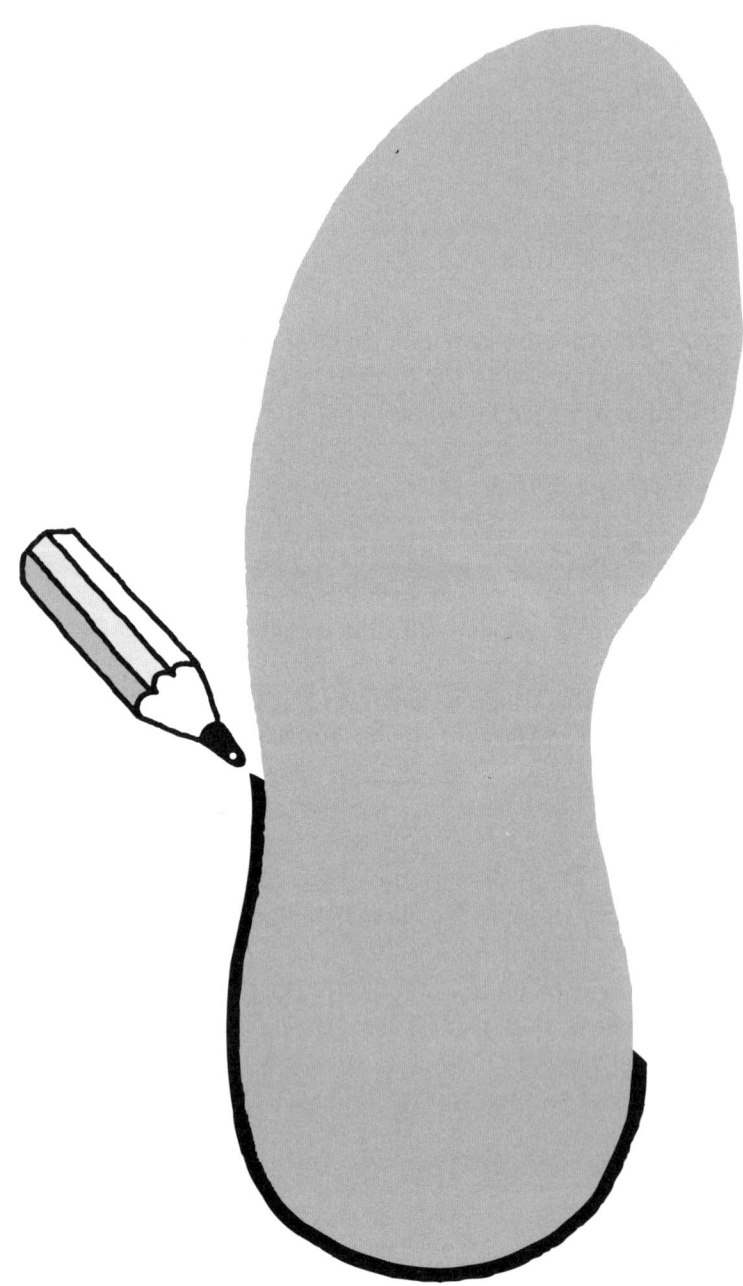

I am _____ footprints tall.

# The Giant's footprint

**Name** _____ **Date** _____

**Work out how tall the Giant is from the size of his footprint.**

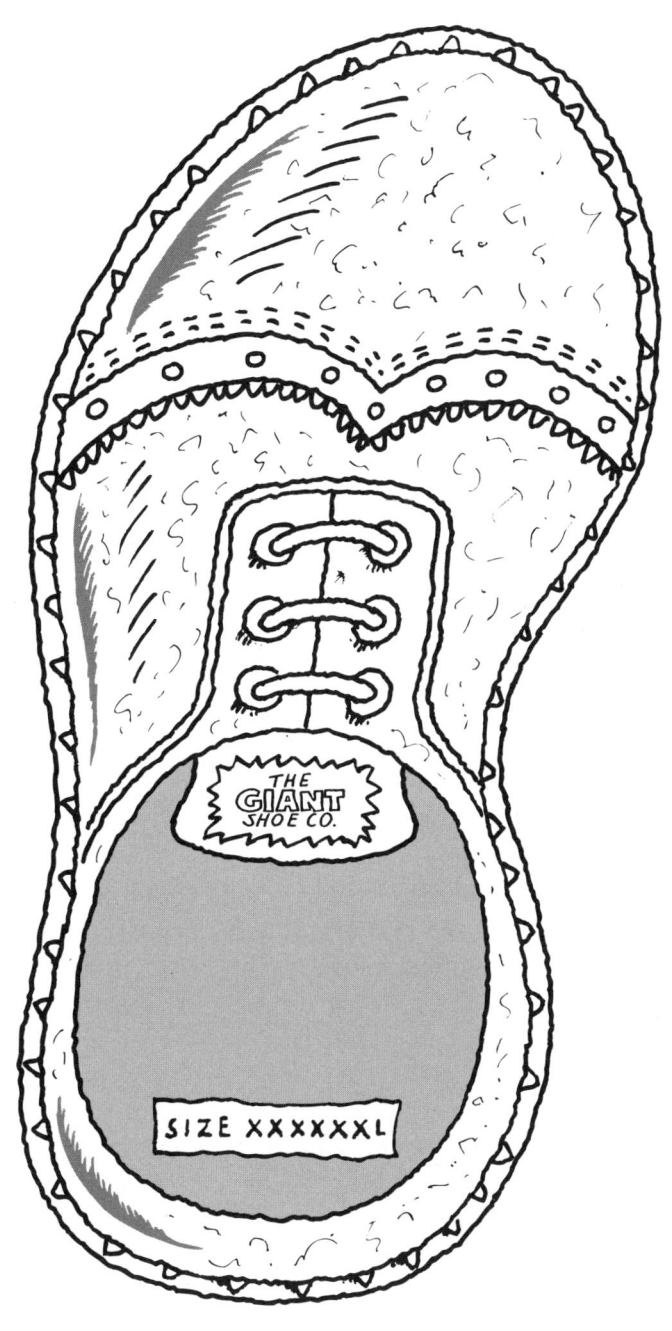

My height is _____ footprints.

The Giant's footprint is _____ .

The Giant's height is _____ .

# Cinderella

In the 'Cinderella' activity the children are presented with a challenging problem that they have to solve. The aim of the activity is to find out the ages of her ugly stepsisters and her stepmother. The game allows the children to apply their creativity as they consider how they will approach helping Cinderella. It is a good opportunity for them to develop their speaking and listening skills and establish links with drama as they create their own sketches.

### Key maths links

- Fractions
- Making decisions
- Rapid recall of multiplication and division facts
- Understanding of multiplication
- Checking results of calculations

### Thinking skills

- Creativity
- Reasoning
- Evaluation
- Problem solving

### Language

double, half, add, subtract, count, predict, guess, too many, too few, enough, altogether, equals How many? How many more to make …?

### Resources

**PCM 15** (one per pair)
**PCM 16** (one per pair)
**copy** of the story of *Cinderella*
**number lines**
**scissors**

 ## Setting the scene

Ask the children if they are familiar with the story of Cinderella. Provide a brief synopsis, if necessary. One quiet evening Cinderella hears the ugly stepsisters and her wicked stepmother talking about their birthdays. She tries to sneak closer and listen but accidentally breaks a vase. She is caught by her stepmother and told that she will have to pay for the vase by selling her precious portrait of her mother (the only thing she has left to remember her mother!) unless she can work out the riddles and find out how old she and the ugly stepsisters are. She has until the morning to work out the answers. The wicked stepmother has added some extra detail to try and confuse Cinderella. The children will need to draw conclusions about who is the eldest, youngest and by how much.

 ## Getting started

Ask the children to work in pairs and give a copy of PCM 15 to each pair. Read the clues, one at a time, to the whole class. They will need to become familiar with the information before they make any decisions and select the ones they work with in order to find out the answers. If necessary, decide as a class where on PCM 16 to record the information from the clues. *How old is Cinderella? What clue helps us?* (younger than 20) The children can now investigate the possible combinations of answers.

### Simplify

The children can spend some time organizing the cards. Guide them into discarding the unhelpful clues and placing the helpful ones in the correct character column on PCM 16. Suggest they begin by considering Cinderella herself (placed in the first column). Help them with their deductions and draw out that Cinderella is between 1 and 19 years old. Suggest they move to the next column and look at the clue placed here. *How much older than Cinderella is Margarita? What must her age range be? Is Margarita older than Lillian?*

### Challenge

The children can try this activity without PCM 16. Suggest that they make their deductions on a sheet of paper. This will give you an opportunity to assess their creativity as they record their workings. *Have you found the useful clues? How will you organize the information you have found? Can you write some extra clues to help Cinderella find out her stepmother's age?*

##  **Checkpoints**

Monitor by sitting with groups of children and observing what they are doing. Only help when necessary. Every so often call the group together and ask them to feedback what they are doing. Share when the children are working well.

### Watch out for ...

Do the children understand all the information?

Ensure the children are sensible and organized in recording what they are saying. Some may be confident in problem solving. Invite them to share their thinking with the class.

### Ask ...

- ● *How are you recording your work?*
- ● *Could they be any other ages? Why not?*
- ● *Who is the eldest?*
- ● *Who is the youngest?*
- ● *How much older is ... than ...?*

### Listen for ...

Observe the children as they work in their groups. Listen for combinations that can work and others that don't, e.g. *Cinderella is between 1 and 19 so Margarita must be between 6 and 24.*

##  **Moving on ...**

Review the solutions found. Invite a few children to share how they solved the problem. Did the layout of PCM 16 help?

### Where next?

- ● Can they make up their own problems?
- ● Could they make up some extra clues to make it easier for other children to solve?
- ● Use drama as a basis for making up problems.

**Did you give enough support when it was needed? What would you change if you repeated the activity? Did any children surprise you? How would you adapt this activity next time?**

DEBRIEF

# Cinderella's problem

**Cut out the cards. Sort out the helpful clues.**

| | | | |
|---|---|---|---|
| age is even number | younger than 20 | 5 years older than Cinderella | twice age of Margarita |
| 2 years older than Margarita | Cinderella's two stepsisters | prettier than Margarita | likes |
| | does not like | the youngest | |

**Thinking by Numbers 1 • Unit 4: What if …? • Creative thinking skills**

# Solving Cinderella's problem

**Find out the ages of Cinderella's ugly stepsisters and her wicked stepmother.**

| Cinderella | Margarita | Lillian | Stepmother |
|---|---|---|---|
| | | | |

## Assessing progress

Assessing the development of creative thinking is challenging as there are often a number of solutions and ideas that can be considered creative in any particular situation. You will have to consider the individual pupil too. A genuinely creative thought for one pupil – something new and insightful for them – may not be so creative in another. Also the process is not necessarily regular or frequent. It is therefore important to consider children's attitudes or their dispositions in different situations. They should be asking questions and confident to offer ideas. It is this confidence or perhaps playfulness that is the best indicator of creativity, rather than trying to assess specific solutions or outcomes.

## Cross-curricular thinking

### Literacy

Brainstorming for ideas is a good general technique to develop creative thinking. It is important that it is done in an atmosphere where the children know that offering ideas is more important than coming up with the right answer and where all ideas are accepted uncritically. In literacy this technique can be used when responding to a text to record thoughts and feelings, as well as to stimulate ideas for composition in terms of the content and detail of the vocabulary used. Brainstorming is usually conducted as a whole class activity. It can also be useful to start off in groups so that the children become more independent in using the technique.

### Science

Using analogies can be a powerful way to develop scientific understanding in a creative way. Asking children to think of an analogy for something (such as an electric current being like water pipes with the current flowing round a circuit) not only provides an opportunity to compare why they are alike and how they are not alike, but also offers an insight into children's thinking about the science involved.

### Design Technology

Coming up with ideas is an essential part of the design process. One technique that can help is to ask children to visualize how the product will be used. Ask them to 'see' it once it is finished: *What will it need to do?, What will make your idea different or special?* This can be the basis for more structured planning and development, though the whole process is a creative one.

### Geography

The strategy 'Banned' (see page 23) can easily be developed in other subjects and provides opportunities to develop specific vocabulary. However, it can also be a way to stimulate creative thinking as children will come up with imaginative ways to give clues to words, such as: *It sounds like fountain but starts with an 'm'.* They may find it hard to formulate rules for banned ideas or words. It is usually best to praise their ingenuity and finish with a discussion of creative ways to get round the rules.

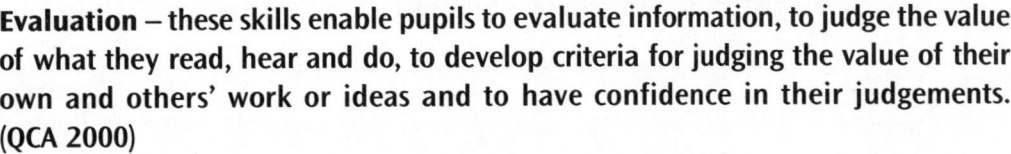

# In my opinion ...

## Evaluation skills

**UNIT 5**

> **Evaluation** – these skills enable pupils to evaluate information, to judge the value of what they read, hear and do, to develop criteria for judging the value of their own and others' work or ideas and to have confidence in their judgements. (QCA 2000)

**Overview**

Evaluation is about taking responsibility for your own opinions and judgements and being prepared to explain or defend them to others with reasons. It requires confidence in knowing what you think and sensitivity in evaluating or criticizing the work of others. It requires the ability to set and apply criteria to tasks and make judgements based on those criteria. The final stage is presenting these judgements to others and being prepared to defend or change that judgement in the light of feedback. This involves awareness of the feelings of others in giving and receiving feedback – a challenging aspect of effective collaboration and an important aspect of speaking and listening.

In mathematics evaluation is essential in developing confidence in knowing that you have a good solution and understanding why. Mathematics is often perceived as being about applying rules or being able to remember facts and formulas. However, an essential part of being able to think mathematically is to be able to make judgements about which facts to use or which formula to apply. A good solution in mathematics might be an efficient one, or an elegant one, or one that leads to new insights and thinking. Deciding which is the best way to do something mathematically therefore, often calls for evaluation and judgement.

**Strategies**

Evaluation skills can be broken down further into the following kinds of behaviours or activities that pupils can do:

- **Evaluate information**
  Appraise, assess, critique, decide
- **Judge the value of what they read, hear and do**
  Review, weigh up, scrutinize
- **Develop criteria for judging the value of their own and others' work or ideas**
  Evaluate, judge, mark
- **Being confident in their judgements**
  Express opinions, disagree, agree (with reasons), resolve

**Questions**

*How could you justify that? What reasons are important? Can you explain ...? How will you check it? Can you argue the opposite? Do you agree? Do you disagree? Which do you think?*

# Aliens

In 'Aliens' the children are presented with a dilemma in that they are told how much a toy costs but are only allowed to pay using silver coins. The children will need to be detectives working with limited information to help them solve this problem. This activity has great links to role-play situations and it would be worthwhile setting up your role-play area as a toyshop to extend opportunities for speaking and listening and problem solving.

## Key maths links

- Estimating and rounding
- Understanding addition and subtraction
- Problems involving 'real life'; money and measures

## Thinking skills

- Evaluating
- Reasoning
- Creative thinking
- Systematic enquiry
- Understanding casual relationships

## Language

counting, addition, subtraction, coin, penny, pence, price, cost, costs more/less, change, total, pay, buy, same, different, list
How much? I think …

## Resources

**PCM 17** (one per pair)
**PCM 18** (one per pair)
**large coins** (5p, 10p, 20p and 50p)
**real coins** (5p, 10p, 20p and 50p)
**whiteboards** (one per pair)
**role-play area** (set up as a toyshop)

 **Setting the scene**

Show the children an enlarged copy of PCM 17 and explain the problem to them. *Ali wants to buy this alien for 30p. He has to use silver coins.* Have copies of large coins available and suggest to the children that they sort out the coins that they can use. *Can Ali pay using a 2p coin? Why not? Can you say which large coins Ali needs to pay for this alien?* (10p +10p+10p) *Are there any other combinations?* Ask the children to think, pair and share their ideas using whiteboards. Take feedback and demonstrate how to complete the problem systematically using PCM 18. *Can anyone think of an easier way of recording? Let's look at the possibilities, e.g. lists, Venn diagram or table.* Give them time to discuss with their partner how they will record their work. *Is everyone clear they know how to record their work?*

 **Getting started**

Ask the children to work in pairs to solve the other problems on PCM 17 and record their answers and workings on PCM 18. Then ask the pairs of children to come back together as a class and look at their solutions. *Is there only one solution? Did anyone have an easy way of recording to help the rest of us?* Share methods, e.g. *This worked well because …*

Show one method of recording the possible silver coins making up 30p from:
20p + 10p; 10p + 10p + 10p; 5p + 5p + 10p + 10p etc.

Move on to the next problem on PCM 17. *How can he pay for the alien costing 40p using silver coins only?*
(20p + 20p; 10p + 10p + 20p; 5p + 5p + 10p + 20p etc.)

### Simplify

Ask the children to use the real coins and to aim for a total of 20p. Discuss how to record their answers; they might want to draw around the coins or use workings/jottings to help them. If necessary, model the use of PCM 18 with the children.

### Challenge

Ask the children to buy an alien costing 45p. *How many ways can you use silver coins to buy this alien? What if the alien costs 50p? How many ways are there now? Is there an easy way of working out the answer or do you need to start from scratch each time?* (Use the previous answer and add to it)

 **Checkpoints**

Some children will not be aware that there is more than one solution; they will need to use trial and error to check their results. *Are you sure you have found all the solutions?*

It is important that you use mini-plenaries to share effective strategies, and develop effective ways of working together. The children also need to be given opportunities to share their methods of recording so they can evaluate and make decisions based on these evaluations.

### Watch out for ...

Praise children who work systematically and show their workings. Model written recording throughout to help those children who struggle. Use drawing the coins and writing the number sentence alongside while modelling. Encourage good listening and explaining

### Ask ...

- *How are you recording your work?*
- *Have you found all the solutions? How do you know?*
- *Did you find a quick way to the answer?*
- *Can we buy something that costs 17p with only silver coins? Why? Why not?*

### Listen for ...

Observe the children working and listen for appropriate vocabulary. Praise children who are making sensible deductions and working well together.

 **Moving on ...**

Review the solutions that the children have worked on. Did they find all the possibilities? How do they know that? Who had an easy way of recording their work? What would you change next time? Was PCM 18 useful? Why? Why not?

### Where next?

- Can the children write their own money problems for someone else to solve?
- Try the same procedure but involving different coins and target totals, e.g. *I have some 10p, 5p and 2p coins. Can I make 27p? What combinations of coins could I have?*
- In the plenary ask the children to evaluate tables etc. Were they useful?

Were the children able to express themselves clearly? Did they find all the solutions? What worked well? Did any of the children surprise you? How clearly were they able to compare methods of recording? Do you need to work on their ability to evaluate? How can you build this into other areas of the curriculum?

DEBRIEF

**Name** _____ **Date** _____

30p only

40p only

20p only

45p only

Use these silver coins to buy these aliens. Use as many as you need.

**Workings**

**Name** _____ **Date** _____

| 30p only | 40p only | 20p only | 45p only |

| 5 | 10 | 20 | 50 | **Total** |
|---|---|---|---|---|
| | | | | |

# Funny fish

**BRIEF**

In 'Funny fish' the children are presented with tropical fish with different numbers of spots. The children are asked to help Layla find out how many fish she has. This activity gives you the opportunity to hear their reasoning and thinking as they justify their ideas. It also gives the children a chance to evaluate how they have worked.

## Key maths links

- ▶ Rapid recall of multiplication and division facts
- ▶ Understanding multiplication and division
- ▶ Making decisions
- ▶ Organizing and using data

## Thinking skills

- ▶ Reasoning
- ▶ Evaluation
- ▶ Problem solving
- ▶ Creative thinking
- ▶ Reflective thinking
- ▶ Using and applying

## Language

count on, adding, altogether, one more, two more, total, more, equals What could I try next? I think ...

## Resources

**PCM 19** (one per pair and one enlarged)
**PCM 20** (one per pupil)
**counters**
**plenary dice**

### Setting the scene

Explain to the class that Layla has a collection of spotted fish. She can count the total number of spots and wants to work out how many of each type of fish she has: how many Bubble Blues and how many Yo-Yo Yellows. Show an enlarged version of PCM 19.

### Getting started

Share PCM 19 with the children. Explain the problem to the class. *Can we help Layla? Her fish have 26 spots. She has 2 types of fish.* The children now need to work out how many of each type of fish Layla has. Demonstrate how to record workings by putting counters on PCM 20 to show a Bubble Blue (3 spots) and a Yo-Yo Yellow (5 spots). *Can you work in pairs to find out how many of each fish there are? Is there only one solution?*

Bring the class together for a plenary so the children have an opportunity to evaluate their learning. Use a plenary (giant) dice with evaluative questions on each face. The pairs take turns to roll the dice and share their responses with the class or another pair.
Use the following evaluative questions on the plenary dice:
*I worked well with ... because ...*
*Today I have learnt ...*
*I liked ...*
*I found ... easy.*
*I found ... hard.*

### Simplify

Share PCM 19 with the children. Change the problem so that Layla's fish have 17 spots.

### Challenge

Share PCM 19 with the children. Change the problem so that Layla's fish have 37 spots.

Suggest the children go on to think about other numbers of spots. *Can you make all the numbers from 8 to 15?*

## 3 Checkpoints

How well have they engaged in the task? Is paired work successful? How well are they counting on? Monitor by sitting with groups of children and observing what they are doing, helping only if necessary.

### Watch out for ...

Applaud the children who know when all possibilities have been found. Praise the children using clear methods of recording, such as jottings. Are the children clear using the language of addition? *What helped you come to a solution?*

### Ask ...

- ◗ *How are you recording your work?*
- ◗ *Have you found all the solutions?*
- ◗ *What were you thinking as you did that?*

### Listen for ...

Observe the children as they talk to each other. Listen for children working together to solve the problem and a way of recording it.

## 4 Moving on ...

Review the combinations to see if they have all been found. Write a class list in a systematic order. How many different ways were there to solve the problem?

### Where next?

- ◗ *I have some triangles and squares in a bag. I can count seven sides in total. How many squares? How many triangles?*
- ◗ *I have some 2p coins and 5p coins in a bag. I can count 9p in total. How many 2p coins? How many 5p coins?*

What worked well? Did you intervene only when really necessary?
How did the children work as pairs? Would you prefer the children to
work as a group? Do you need to do more work on these objectives?
Were there any surprises?

DEBRIEF

# Funny fish

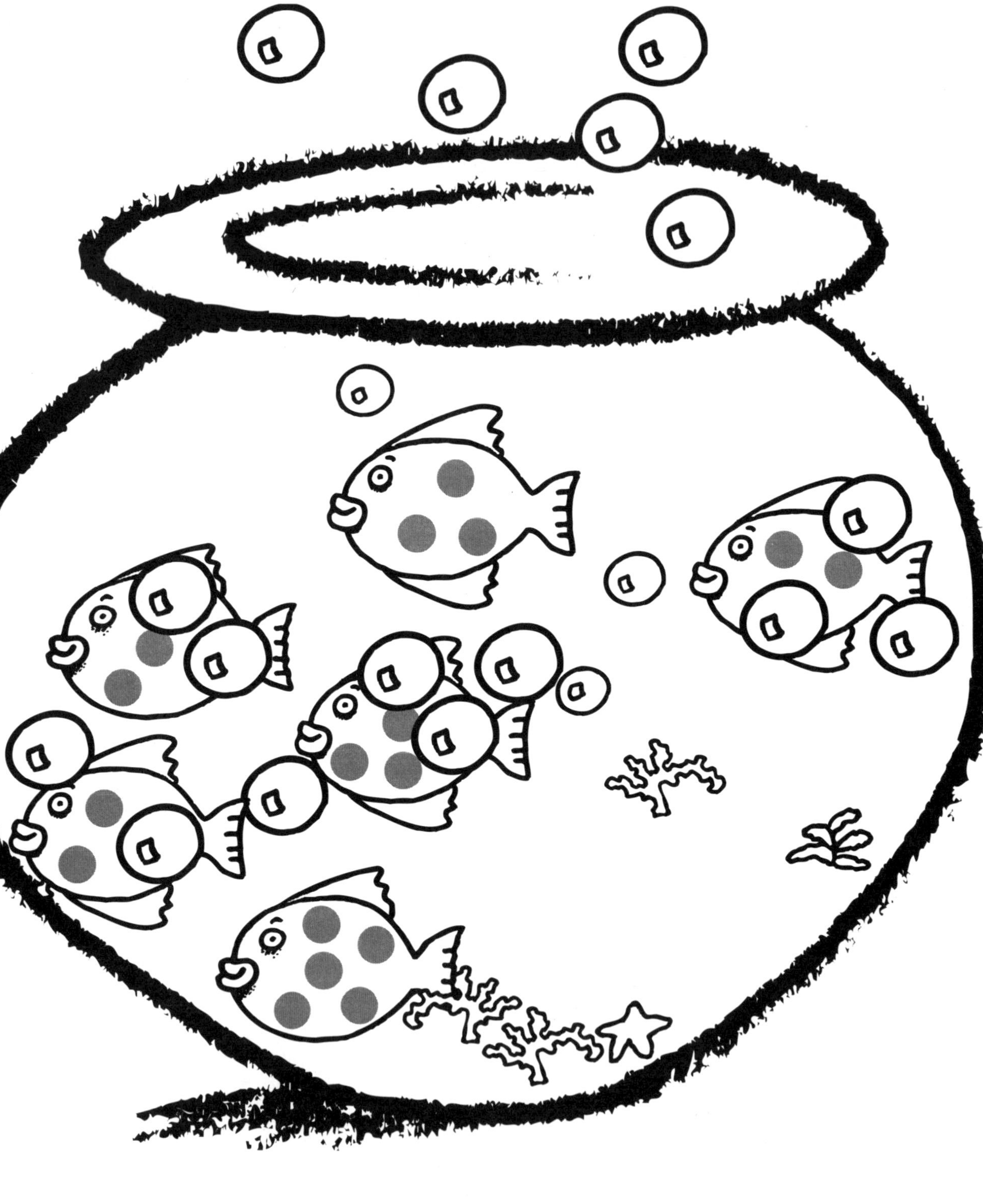

Bubble Blues have 3 spots. Yo-Yo Yellows have 5 spots.

PCM 20

# Bubble Blues and Yo-Yo Yellows

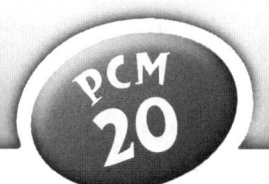

**Name** ______________________________ **Date** ____________

How many Bubble Blues and Yo-Yo Yellows are there in Layla's tank?

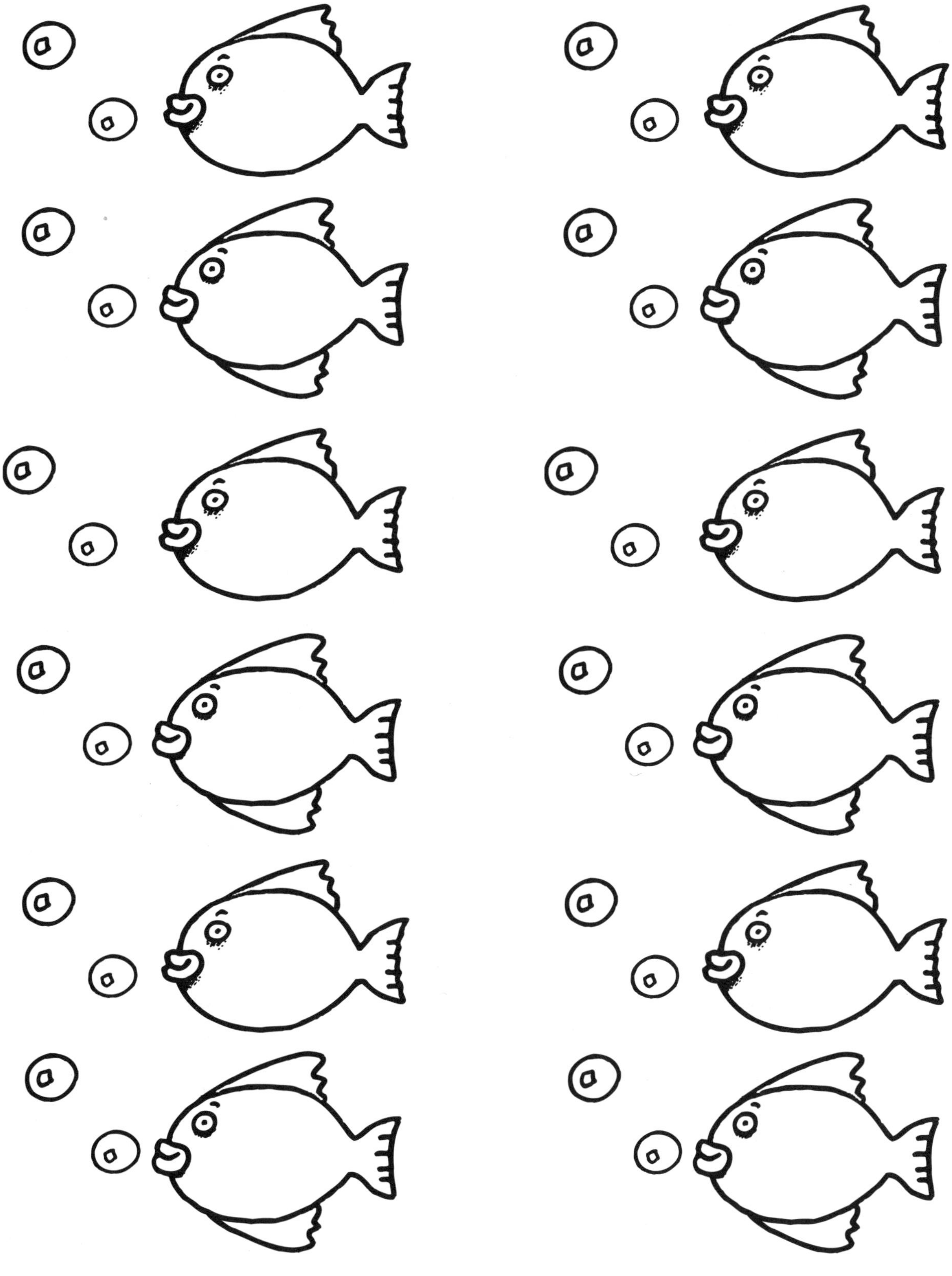

## Assessing progress

The children's increased confidence in their own thinking is one of the hallmarks of improving evaluation skills. It is about taking responsibility for your own opinions and judgements and being prepared to explain or defend them to others with reasons. This requires confidence in knowing what you think. This confidence should be justified, of course, so children should be prepared to change their minds if necessary, in the light of information or reasoning. At this stage it is also important for children to show sensitivity in evaluating or criticizing the work of others.

## Cross-curricular thinking

### Literacy

Assessment for learning (see page 14) strategies such as 'Traffic lights' are good starting points to develop evaluation skills. You can ask the children to rate a piece of writing that they have done with green for: *I think I can go on*, orange for: *I think I am getting going*, and red for: *I'm at a full stop here*. This opens up the way to discuss criteria for success in the task so that children can evaluate their own work.

### Design technology

Evaluation is also central to design technology. The children need to learn to develop evaluation criteria for their designs in order to guide their thinking as they work. This should be an integral part of the process and not simply a retrospective review. Using a digital camera to record the process of designing and making enables the children to recall what they were thinking at the different stages and reflect on the criteria to evaluate the task.

### History

A strategy such as a 'Mystery' (where snippets of information are pieced together to answer a central question) can help children to use their evaluative skills as they judge the importance of the different 'clues' they have been given. In history this can be a good way to assess understanding of what has been learned in a unit of work as they use their historical knowledge to do this. Clues can easily be written to support a discussion about: *Who was responsible for the Great Fire of London?* for example, to get children to see that the baker may have started the fire, but that there are other factors to consider.

### Geography

Some other general techniques that are helpful in developing evaluation skills are those developed by Edward deBono where children are given thinking frames with headings such as 'Plus, minus and interesting' (PMI) or a focus on 'Consider all factors' (CAF). The structure of the sheet helps children to think more carefully and give more considered responses. These approaches can be combined with collaborative discussion (such as 'Think, Ink, Pair, Share' where children are asked to consider their response, make some notes, discuss it with a partner then in a group). This can be particularly useful in a subject like geography when the children have to evaluate changes to the environment or express their views about people and places.

# Think on!

## Using and applying thinking skills

> **Using and applying thinking** – in mathematics these skills involve pupils in developing the skills and strategies that will help them solve problems they face both in learning at school and in life more broadly. They involve problem solving in its broadest sense and include the skills of identifying and understanding what the issue or the problem is, planning solutions, monitoring progress in tackling the issue or problem and then reviewing and evaluating any solutions.

**Overview**

The aim of this unit is to identify some activities for pupils to put their mathematical thinking skills into practice. This will give them the opportunity to evaluate how well they have developed their skills through the earlier activities as well as giving you the opportunity to assess how well they can apply what they have learned. The activities are set as challenges, problems or puzzles.

The process of undertaking these activities relates to the different kinds of thinking in the earlier units. The early stages draw on information processing skills by focusing the children on what they have to do and what they already know. There may be scope for creativity in seeing alternatives or applying knowledge and skills imaginatively to a new problem. Enquiry skills are brought into play during the main part of the activity as any solution is formulated and tested, closely supported by reasoning skills which also help to link the different stages and ensure continuity throughout the process. Evaluation skills are essential to appraise and review any solution and to develop confidence in being successful.

**Strategies**

Supporting the pupils in using and applying thinking skills to problems is best framed as a series of questions:

- **What do we have to do?**
  *What is the problem, challenge or issue to be resolved?*
- **Where do we start?**
  *What do we know?*
  *Have we done anything like this before?*
  *What possibilities are there?*
- **How will we know when we have got there?**
  *What will a successful solution look like?*
- **Are we on track?**
  *Is this going to lead us to the answer we imagined?*
- **Have we got there?**
  *Is this a solution to the problem we were set?*
  *Could we have done it differently? Is it the best solution?*

**Questions**

*What do you have to do? What do you need to know? What do you know already? Have you seen anything like this before? What could you try? Do you think that will work? What will the answer look like? How could you test that? How can you check that? Is this the best answer? How else could you have done it?*

# Treasure hunt

## BRIEF

In 'Treasure hunt' the children can participate in an activity that develops links with speaking and listening and drama. They can imagine that they are pirates searching for treasure, and if they have maps that they have painted in art they can use them to follow imaginary clues to the golden coins. Once the treasure is discovered the children will then need to apply their knowledge of odd numbers to the problem of sharing it out fairly/unfairly.

## Key maths links

- Place value and ordering
- Rapid recall of addition and subtraction facts
- Problems involving 'real life'

## Thinking skills

- Reasoning
- Systematic enquiry
- Problem-solving
- Creative thinking
- Reflective thinking
- Using and applying

## Language

counting, addition, subtraction, odd, even, coin, penny, pence, price, cost, costs more/less, change, total, pay, buy, same, different, list
How much? I think …

## Resources

**PCM 21** (one per group)
**PCM 22** (one per pupil)
**whiteboards**
**role-play area** (set up as pirate's ship)
**coins or counters** (21 per pupil)
**treasure maps** (pupils' own)
**happy/sad cards** (three)

## Setting the scene

Explain that three pirates, Blue Beard, Black Beard and Long-Jack Silver, have been working together to locate some missing treasure. They have found 21 golden coins to put in their treasure chests. Unfortunately, they can only put an odd number of coins in each chest. The children need to explore the possible combinations of sharing the coins amongst the pirates. *Is there a fair way to share the coins?* (7 in each) *Which is the fairest way? Which is the worst (most unfair) way for the pirates?*

## Getting started

Give out PCM 21 and read the problem together. Suggest that they work in groups of three. Encourage the children to use PCM 22 to record their ideas. *How many ways can the pirates share the 21 coins?* (12 ways)

Model the start of a systematic way of recording all the possibilities and encourage the children to continue themselves. Remember to reinforce that 1, 1, 19 is the same as 19, 1, 1 etc.

Bring the children together to share their ideas. Did they find all the ways? How can they check? When you are sure of their understanding they can apply their ways of working to the sharing of 23 coins. (14 ways)

During the plenary the children will need a chance to think about fair/unfair ways of sharing. On three cards draw a happy face on one side and a sad face on the other. Invite three children to stand up and tell them the amount they have in their chests is 1, 1, 19 respectively. They indicate how they feel by showing the appropriate side of the card. *Who is happy? Who is sad? Why? Is it fair? Can we find a way of making all the pirates happy? Which is the most unfair?*

### Simplify

The children can to use counters to place in the different chests on PCM 22. *What is an odd number?* Demonstrate by sorting the counters into two piles. *Odd numbers don't go into two piles with the same number in each pile.* If they are not confident with working with numbers to 21 they should start by sharing 11 coins. (4 ways)

### Challenge

Once the children have tackled and completed the problem on PCM 21 develop their thinking by encouraging them to apply their knowledge through a brainstorming and questioning session. *What would happen if another pirate joined them? Can you adapt PCM 22 to show your workings? How many coins would they need to make it fair?* (any multiples of 4: 4, 8, 12, 16, 20, 24, 28) *Why?*

**Checkpoints**

How are they recording their work? Are they working well in their groups of three? Do they need counters/counting equipment to aid their calculations? Are all the groups confident with adding to 21?

### Watch out for ...

Praise systematic ways of working and showing workings using PCM 22. Focus on the common difficulties and identify children who quickly find a solution. Undertake to model written recordings throughout to help those children who struggle. Use drawing the coins and writing the number sentences alongside as one method of recording.

### Ask ...

- ❍ *What is an odd number?*
- ❍ *How do you know that?*
- ❍ *Why do you think that?*
- ❍ *Can you find another way?*
- ❍ *What if ...?*
- ❍ *What happens when you add two odd numbers?*
- ❍ *What is fair? Unfair?*

### Listen for ...

Encourage use of appropriate vocabulary, e.g. sharing, total, adding, odd. Look for children adding three numbers confidently. Identify groups working well together.

 **Moving on ...**

Review the solutions found. Did they find all the possible combinations? *What helped you find the right answer? Who worked well together? Why? What is an odd number?*

### Where next?

- ❍ Can the children write their own money problems for someone else to solve?
- ❍ *What happens when you add two odd numbers?*
- ❍ *What happens when you add two even numbers?*
- ❍ *What would happen if another pirate joined them? Can you share 21 coins between them fairly?*
- ❍ *What would happen if you could only put even numbers of coins in the treasure chests? Can you start with an odd total?*
- ❍ Following instructions and directions on maps links with geography subjects.

**Do you need to do more work on money through role-play? How can you improve this task for next time? Did anyone surprise you? Was the activity supportive enough for the less able?**

DEBRIEF

**Name** _____ **Date** _____

Three pirates found 21 golden coins. Can they share the coins fairly?

What would happen if they found 23 coins?

**Name** _____   **Date** _____

How many coins in each chest?

Remember odd numbers only.

# Beautiful butterflies

**BRIEF**

In 'Beautiful butterflies' the children are presented with butterflies with wings that have space for three colours on each wing. The children are challenged with the task of seeing how many different butterflies they can make. They will need to apply their knowledge of symmetry to this activity. It also develops the children's recording skills as they will need to organize the information clearly. Each butterfly has to have three colours (two of each). This activity has useful links to art and it is a great way of reinforcing symmetry, especially if the children have completed butterfly paintings in an art lesson.

## Key maths links

- Reasoning about numbers or shapes
- Shape and space

## Thinking skills

- Reasoning
- Systematic enquiry
- Problem-solving
- Creative thinking
- Reflective thinking
- Using and applying

## Language

fold, match, mirror line, reflection, symmetrical, line of symmetry, half

## Resources

**PCM 23** (one per group)
**PCM 24** (one per group)
**mirrors**
**pictures of butterflies**
**symmetrical butterfly paintings** (pupils' own) (paint – then fold and rub together)
**coloured pencils**
**sticky paper squares**

## Setting the scene

What do they know about butterflies? Prompt for the symmetrical properties. Show pictures of butterflies to reinforce this and refer to their previous work completed in art lessons (butterfly paintings).

Show an enlarged version of PCM 23. Explain that the butterflies have to be symmetrical. Demonstrate a symmetrical example, e.g. red, blue, green, green, blue, and red. Then colour red, blue, green, red, blue, green. *Is this symmetrical? Why not?* Use a mirror to reinforce symmetry. The children need to be confident working with symmetry so you will need to make explicit the equivalent sections on the wings that are required for the butterflies to be symmetrical, e.g. *If I colour this section red, what will be red on this wing? Can anyone show us?*

## Getting started

Once you are confident that the children clearly understand the term symmetrical then let them have PCM 23. Ask them to work in pairs as they tackle the task. They will need different coloured pencils (red, blue and green). Reinforce that each butterfly will have six sections: two red sections, two blue sections and two green sections. *How many different butterflies can you make?* (6 different butterflies)

Bring the class together to share all the possible butterflies. Check the children understand all the possibilities before giving them another task on PCM 24. *What if the butterfly has eight sections?* Again this would give two sections of each colour, e.g. two sections of green, two of red, two of blue and two of yellow. *How many different butterflies can you make now?* (24)

### Simplify

*Can you make symmetrical butterflies?* Work with the children and establish that they are confident with the rules of symmetry. Encourage the group to use mirrors to check their work. *How many different symmetrical butterflies can you make?* The children will need to work closely as a group to check that each butterfly is symmetrical and different.

### Challenge

After the children have successfully completed PCM 23 and PCM 24 set a different task. Tell them that the butterfly has six sections and they only have three colours (any colours) but they can colour two or three sections the same. *How many different butterflies can you make now?* (9)

If the task is extended to colouring one, two or three sections the same (using only three colours) then there are 27 different butterflies.

 ## Checkpoints

Every so often, call the group or class together and ask them to feedback on what they have done so far. Is everyone clear on what makes something symmetrical? If the children are finding this activity challenging then get them to experiment with sticky paper squares. They can then fold the halves together and check they are mirror images.

### Watch out for ...

Some children may need support with understanding symmetry; they will need mirrors to check their work. Use mini-plenaries to share their understanding of the task and to correct any misconceptions.

 ### Ask ...

- *Have you found all the solutions? How do you know?*
- *Is that butterfly symmetrical?*
- *Can you predict how many different butterflies you will find?*
- *What if we only had two sections and two colours?*

 ### Listen for ...

Listen for the use of appropriate vocabulary, e.g. same, symmetrical. Praise children who are working well together. Listen for confident explanations, e.g. *I know this is symmetrical as I used my mirror to check.*

 ## Moving on ...

Review the solutions found. *Are they all different? Are they all symmetrical? Did you find more with four colours and eight sections?* Check their understanding of symmetry by asking them to get into pairs and make a symmetrical shape with their partner. Look at the shapes they have made: *Is this symmetrical? How do you know?*

### Where next?

- Look at symmetrical patterns in art.
- Use cubes or sticky coloured squares to create symmetrical shapes and patterns.
- Can you complete this pattern?

**Do you need to do more work on shape and space in the numeracy lesson? Did the children have enough structured support? Did anyone surprise you?**

DEBRIEF

# My 3-colour butterflies

**Name** _____  **Date** _____

Your butterfly has 6 sections:
2 red 2 blue 2 green.

How many different
butterflies can you make?

Remember to
use 3 colours
on each wing.

# My 4-colour butterflies

**Name** _____  **Date** _____

Your butterfly has 8 sections: 2 red 2 blue 2 green and 2 yellow.

How many different butterflies can you make? Are they all different?

Remember to use 4 colours on each wing.

# The growing caterpillar

**BRIEF**

In 'The growing caterpillar' the children are presented with simple graphs that have mistakes. They need to apply their knowledge and find the mistakes and correct them as a class. The children are then asked to make their own graphs based on the stories they tell. This activity gives you the opportunity to hear their reasoning and thinking as they justify their ideas. It also gives them a chance to apply their knowledge of data handling to identify mistakes on the graphs and make their own graphs. This activity develops cross-curricular links with non-fiction texts, such as life cycles in science.

## Key maths links

- Making decisions
- Organizing and using data

## Thinking skills

- Reasoning
- Systematic enquiry
- Problem solving
- Creative thinking
- Reflective thinking
- Using and applying

## Language

more than, less than, favourite, predict, Monday, Tuesday, Wednesday, Thursday, Friday, Saturday
I think …, What could I try next? How many more to make …? I know …

## Resources

PCM 25 (one per pair and enlarged)
PCM 26 (one per table)
copy of *The Very Hungry Caterpillar* by Eric Carle
story sack
scissors

 ## Setting the scene

Read *The Very Hungry Caterpillar* by Eric Carle. Encourage the class to join in. Now look at a graph (that is incorrect) made from PCM 25 and PCM 26. Show only Monday to Thursday. Explain to the children that the graph shows all the things a growing caterpillar ate on Monday, Tuesday, Wednesday and Thursday. Unfortunately though, there are some mistakes. *Can we help?*

## Getting started

Share and go through the graph with the class and retell the story together. *What did he eat on Monday? How many apples do we have? What did he eat on Wednesday? Can you spot any other mistakes?* Work in pairs to spot any mistakes. Take off/add tiles until your graph is correct. *On which day did he eat the most? What would happen if he ate two more apples on Monday? How many would we need now?*

Explain that in their groups the children are going to make up their own versions of the story. Can they make a graph to show their story? They will need prompting about working together and taking turns. Give them copies of both PCMs (only need section of graph to show Monday to Thursday) to create their graphs. Explain that at the end of the lesson each group will show their graph and tell their story. Use mini-plenaries to check the understanding and ability of the groups to work together. Can they write some questions to ask the other groups about their graph? *How many … did he eat on Monday? On what day did he eat the most?*

### Simplify

Use a story sack with the children to retell the story so that they are familiar with the information. Recreate the graph using the story from the table (only need section of graph to show Monday to Thursday). *On what day did he eat the most? Least?* Let them add to/take away from this graph. They then work as a group to retell the story. This group will need support to develop their confidence. They then need the opportunity to create a new graph and retell the story.

### Challenge

Use the whole graph from Monday to Saturday to develop their thinking. *Why is Saturday different? Can you add this information to the graph? Is there a better way of showing what the caterpillar ate?* Once they are happy with how they have presented their information they can write questions to ask the class, e.g. *How many more oranges than apples did he eat? On what day did he eat the most?*

## Checkpoints

How well have they engaged in the task? Is paired work successful? How well are they coping with the task? Monitor by sitting with groups of children and observing what they are doing, helping only if necessary.

### Watch out for ...

Praise children using systematic approaches. Are the children clear when using data handling language? *What helped you come to a solution?*

### Ask ...

- ❍ *How are you recording your work?*
- ❍ *What were you thinking as you did that?*
- ❍ *Did it help working as a table?*

### Listen for ...

Observe the children as they talk to each other. Are they clear about the graph telling us a story?

## Moving on ...

Review the graphs and the new stories. Can you answer the questions the children have written? *Can you create a human graph for your information? What would happen if ...?* (e.g. he ate 10 more apples? How many now?) *On Monday he ate one orange, on Tuesday he ate three apples etc.* As you give the information to the children they make the graph. Are all groups successful?

### Where next?

- ❍ Using the idea of a graph to tell a story, show the chores Cinderella carries out in a week.
- ❍ Use a Venn diagram to develop data handling further through sorting shapes, money, numbers, e.g. coins under 10p/silver coins
- ❍ *If an apple costs twice as much as a pear and a pear cost 10p, how much does an apple cost?*
- ❍ Use labels to price the items, e.g. orange = 5p, apple = 10p etc. *How much did the caterpillar spend?*

What worked well? Did you intervene only when really necessary?
How did the children work in groups? Would you prefer the children to
work in groups or pairs? Do you need to do more work on these objectives?
Were there any surprises?

DEBRIEF

## What did the growing caterpillar eat?

**Living graph**

Monday  Tuesday  Wednesday  Thursday  Friday  Saturday

5  4  3  2  1

PCM 26

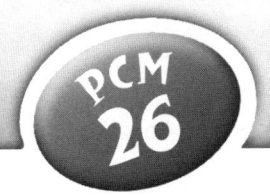

# Caterpillar food

## Cut out and use on the graph.

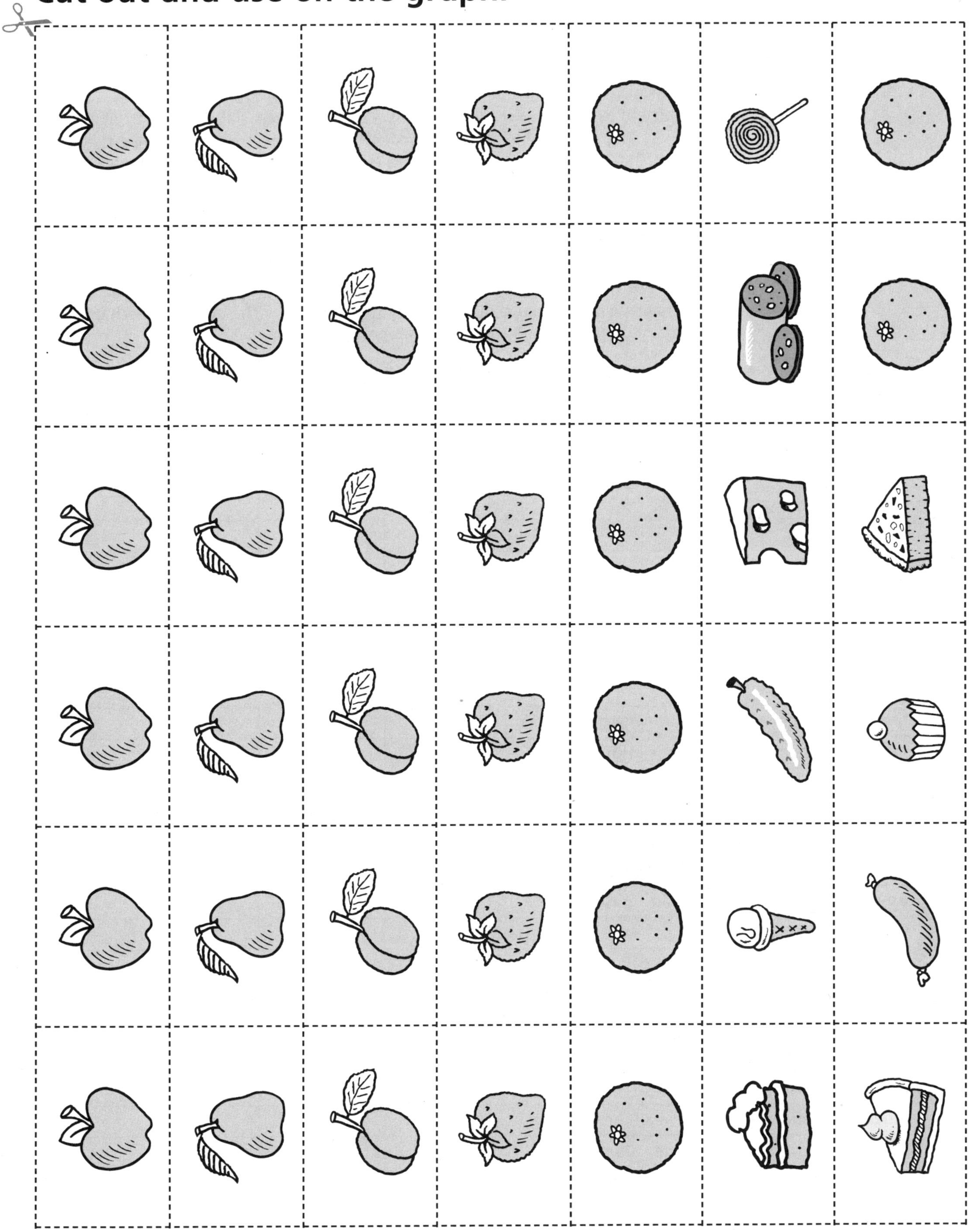

## Assessing progress

The aim of this final unit was to offer some activities for pupils to put their mathematical thinking skills into practice. This should have given you the opportunity to evaluate how well they have developed their skills through the earlier activities as well as the opportunity to assess how well they can apply what they have learned. The grid below is a way for you to review where you think the children have made progress. It is designed for you to use on the whole class, but could be used to reflect on individual children. It is set out as a grid so that you can indicate where you think the first five units were successful, whether the children were able to show these skills in the activities in Unit 6, where you think you have seen progress in other areas of the curriculum, and where you think the children have developed their awareness of their thinking skills. You may wish to review the activities with a colleague who has also been using the *Thinking by Numbers* activities.

| Thinking skills | | Units 1–5 | Unit 6, Using and applying | Across the curriculum | Awareness of the skills |
|---|---|---|---|---|---|
| Information processing | locate and collect relevant information | | | | |
| | sort | | | | |
| | classify | | | | |
| | sequence | | | | |
| | compare and contrast | | | | |
| | analyse part/whole relationships | | | | |
| Reasoning | give reasons for opinions and actions | | | | |
| | draw inferences | | | | |
| | make deductions | | | | |
| | use precise language to explain what they think | | | | |
| | make judgements and decisions informed by reasons or evidence | | | | |
| Enquiry | ask relevant questions | | | | |
| | pose and define problems | | | | |
| | plan what to do and how to research | | | | |
| | predict outcomes and anticipate consequences | | | | |
| | test conclusions | | | | |
| | improve ideas | | | | |
| Creative thinking | generate and extend ideas | | | | |
| | suggest hypotheses, to apply imagination | | | | |
| | look for alternative innovative outcomes | | | | |
| Evaluation | evaluate information | | | | |
| | judge the value of what they read, hear and do | | | | |
| | develop criteria for judging the value of their own and others' work or ideas | | | | |
| | have confidence in their judgements | | | | |

# Appendix

## Scope and sequence chart

| Unit | Unit name | Activity name | Key maths links | Thinking skills | Page no. |
|------|-----------|---------------|-----------------|-----------------|----------|
| 1 | Sort it out! *Information processing skills* | Little Red Riding Hood | ⊙ Solve mathematical problems or puzzles<br>⊙ Recognize turns to the left or to the right<br>⊙ Give instructions for moving along a route | ⊙ Sorting<br>⊙ Find relevant information<br>⊙ Comparing<br>⊙ Contrasting | 26–29 |
|  |  | The Bears' breakfast | ⊙ Place value and ordering<br>⊙ Understanding addition and subtraction<br>⊙ Problems involving 'real life'; money or measures<br>⊙ Mental calculation strategies | ⊙ Sorting<br>⊙ Ordering<br>⊙ Comparing | 30–33 |
| 2 | That's because … *Reasoning skills* | Journey to the Moon | ⊙ Counting, properties of number and number sequences<br>⊙ Place value and ordering<br>⊙ Checking results of calculations | ⊙ Giving reasons<br>⊙ Working systematically<br>⊙ Describing | 36–39 |
|  |  | What's in the ring? | ⊙ Reasoning about numbers and shape<br>⊙ Counting, properties of numbers and number sequences<br>⊙ Sort shapes and describe some of their features<br>⊙ Organizing and using data | ⊙ Recognizing properties of numbers and shapes<br>⊙ Classifying numbers and shapes<br>⊙ Identifying similarities | 40–43 |
| 3 | Detective work *Enquiry skills* | How many shapes? | ⊙ Reasoning about numbers or shape<br>⊙ Shape and space | ⊙ Giving reasons<br>⊙ Explaining<br>⊙ Making connections<br>⊙ Working systematically | 46–49 |
|  |  | Hoopla | ⊙ Estimating and rounding<br>⊙ Rapid recall of addition and subtraction facts<br>⊙ Making decisions | ⊙ Investigating<br>⊙ Problem-solving<br>⊙ Proposing ideas<br>⊙ Testing solutions | 50–53 |
| 4 | What if …? *Creative thinking skills* | The Giant's footprint | ⊙ Estimating and rounding<br>⊙ Measures<br>⊙ Problems involving 'real life'; money and measures | ⊙ Think up ideas<br>⊙ Invent solutions<br>⊙ Imagine possibilities | 56–59 |
|  |  | Cinderella | ⊙ Fractions<br>⊙ Making decisions<br>⊙ Rapid recall of multiplication and division facts<br>⊙ Understanding of multiplication<br>⊙ Checking results of calculations | ⊙ Creativity<br>⊙ Reasoning<br>⊙ Evaluation<br>⊙ Problem solving | 60–63 |

| Unit | Unit name | Activity name | Key maths links | Thinking skills | Page no. |
|---|---|---|---|---|---|
| 5 | In my opinion … *Evaluation skills* | Aliens | ● Estimating and rounding<br>● Understanding addition and subtraction<br>● Problems involving 'real life'; money and measures | ● Evaluating<br>● Reasoning<br>● Creative thinking<br>● Systematic enquiry<br>● Understanding casual relationships | 66–69 |
| | | Funny fish | ● Rapid recall of multiplication and division facts<br>● Understanding multiplication and division<br>● Making decisions<br>● Organizing and using data | ● Reasoning<br>● Evaluation<br>● Problem solving<br>● Creative thinking<br>● Reflective thinking<br>● Using and applying | 70–73 |
| 6 | Think on! *Using and applying thinking skills* | Treasure hunt | ● Place value and ordering<br>● Rapid recall of addition and subtraction facts<br>● Problems involving 'real life', money or measures | ● Reasoning<br>● Systematic enquiry<br>● Problem-solving<br>● Creative thinking<br>● Reflective thinking<br>● Using and applying | 76–79 |
| | | Beautiful butterflies | ● Reasoning about numbers or shapes<br>● Shape and space | ● Reasoning<br>● Systematic enquiry<br>● Problem-solving<br>● Creative thinking<br>● Reflective thinking<br>● Using and applying | 80–83 |
| | | The growing caterpillar | ● Making decisions<br>● Organizing and using data | ● Reasoning<br>● Systematic enquiry<br>● Problem solving<br>● Creative thinking<br>● Reflective thinking<br>● Using and applying | 84–87 |

# Thinking by Numbers 1 and the NNS Medium-term Plans

The following chart shows how the thinking activities could be used if following the teaching order suggested in the NNS Medium-term Plans. Choose an appropriate activity to suit your class.

| Autumn Term | | | | |
|---|---|---|---|---|
| | | **Thinking by Numbers** | | |
| **Unit** | **Unit topic** | **Activity name** | **Thinking skill** | **Page no.** |
| 1 | Counting, properties of numbers and number sequences | Unit 2: Journey to the Moon | Reasoning | 36–39 |
| | | Unit 2: What's in the ring? | Reasoning | 40–43 |
| | | Unit 4: Cinderella | Creative thinking | 60–63 |
| 2–4 | Place value, ordering, estimating, rounding | Unit 1: The Bear's breakfast | Information processing | 30–33 |
| | | Unit 2: Journey to the Moon | Reasoning | 36–39 |
| | | Unit 6: Treasure hunt | Using and applying | 76–79 |
| | Understanding + and – | Unit 1: The Bear's breakfast | Information processing | 30–33 |
| | | Unit 2: Journey to the Moon | Reasoning | 36–39 |
| | | Unit 5: Aliens | Evaluation | 66–-69 |
| | Mental calculation strategies (+ and –) | Unit 1: The Bear's breakfast | Information processing | 30–33 |
| | Money and 'real life' problems | Unit 1: The Bear's breakfast | Information processing | 30–33 |
| | | Unit 4: The Giant's footprint | Creative thinking | 56–59 |
| | | Unit 5: Aliens | Evaluation | 66–69 |
| | | Unit 6: Treasure hunt | Using and applying | 76–79 |
| | Making decisions, checking results | Unit 3: Hoopla | Enquiry | 50–53 |
| | | Unit 4: Cinderella | Creative thinking | 60–63 |
| | | Unit 5: Funny fish | Evaluation | 70–73 |
| | | Unit 6: The growing caterpillar | Using and applying | 84–87 |
| 5–6 | Measures, including problems | Unit 4: The Giant's footprint | Creative thinking | 56–59 |
| | Shape and space Reasoning about shapes | Unit 1: Little Red Riding Hood | Information processing | 26–29 |
| | | Unit 2: What's in the ring? | Reasoning | 40–43 |
| | | Unit 3: How many shapes? | Enquiry | 46–49 |
| | | Unit 6: Beautiful butterflies | Using and applying | 80–83 |
| 7 | **Assess and review** | | | |
| 8 | Counting, properties of numbers and number sequences Reasoning about numbers | Unit 2: Journey to the Moon | Reasoning | 36–39 |
| | | Unit 2: What's in the ring? | Reasoning | 40–43 |
| | | Unit 4: Cinderella | Creative thinking | 60–63 |
| 9–11 | Place value, ordering, estimating, rounding | Unit 1: The Bear's breakfast | Information processing | 30–33 |
| | | Unit 2: Journey to the Moon | Reasoning | 36–39 |
| | | Unit 3: Hoopla | Enquiry | 50–53 |
| | | Unit 4: The Giant's footprint | Creative thinking | 56–59 |
| | | Unit 5: Aliens | Evaluation | 66–69 |
| | | Unit 6: Treasure hunt | Using and applying | 76–79 |
| | Understanding + and – | Unit 1: The Bear's breakfast | Information processing | 30–33 |
| | | Unit 2: Journey to the Moon | Reasoning | 36–39 |
| | | Unit 5: Aliens | Evaluation | 66–69 |
| | Mental calculation strategies (+ and –) | Unit 1: The Bear's breakfast | Information processing | 30–33 |
| | Money and 'real life' problems | Unit 1: The Bear's breakfast | Information processing | 30–33 |
| | | Unit 4: The Giant's footprint | Creative thinking | 56–59 |
| | | Unit 5: Aliens | Evaluation | 66–69 |
| | | Unit 6: Treasure hunt | Using and applying | 76–79 |
| | Making decisions | Unit 3: Hoopla | Enquiry | 50–53 |
| | | Unit 4: Cinderella | Creative thinking | 60–63 |
| | | Unit 5: Funny fish | Evaluation | 70–73 |
| | | Unit 6: The growing caterpillar | Using and applying | 84–87 |
| 12–13 | Measures, and time, including problems | Unit 4: The Giant's footprint | Creative thinking | 56–59 |
| | Handling data | Unit 2: What's in the ring? | Reasoning | 40–43 |
| | | Unit 5: Funny fish | Evaluation | 70–73 |
| | | Unit 6: The growing caterpillar | Using and applying | 84–87 |
| 14 | **Assess and review** | | | |

## Spring Term

| Unit | Unit topic | Activity name | Thinking skill | Page no. |
|---|---|---|---|---|
| | | **Thinking by Numbers** | | |
| | | Activity name | Thinking skill | Page no. |
| 1 | Counting, properties of numbers and number sequences | Unit 2: Journey to the Moon | Reasoning | 36–39 |
| | | Unit 2: What's in the ring? | Reasoning | 40–43 |
| | | Unit 4: Cinderella | Creative thinking | 60–63 |
| 2–4 | Place value, ordering, estimating, rounding | Unit 1: The Bear's breakfast | Information processing | 30–33 |
| | | Unit 2: Journey to the Moon | Reasoning | 36–39 |
| | | Unit 6: Treasure hunt | Using and applying | 76–79 |
| | Understanding + and – | Unit 1: The Bear's breakfast | Information processing | 30–33 |
| | | Unit 2: Journey to the Moon | Reasoning | 36–39 |
| | | Unit 5: Aliens | Evaluation | 66–69 |
| | Mental calculation strategies (+ and –) | Unit 1: The Bear's breakfast | Information processing | 30–33 |
| | Money and 'real life' problems | Unit 1: The Bear's breakfast | Information processing | 30–33 |
| | | Unit 4: The Giant's footprint | Creative thinking | 56–59 |
| | | Unit 5: Aliens | Evaluation | 66–69 |
| | | Unit 6: Treasure hunt | Using and applying | 76–79 |
| | Making decisions | Unit 3: Hoopla | Enquiry | 50–53 |
| | | Unit 4: Cinderella | Creative thinking | 60–63 |
| | | Unit 5: Funny fish | Evaluation | 70–73 |
| | | Unit 6: The growing caterpillar | Using and applying | 84–87 |
| 5–6 | Measures, including problems | Unit 4: The Giant's footprint | Creative thinking | 56–59 |
| | Shape and space | Unit 1: Little Red Riding Hood | Information processing | 26–29 |
| | Reasoning about shapes | Unit 2: What's in the ring? | Reasoning | 40–43 |
| | | Unit 3: How many shapes? | Enquiry | 46–49 |
| | | Unit 6: Beautiful butterflies | Using and applying | 80–83 |
| 7 | **Assess and review** | | | |
| 8 | Counting, properties of numbers and number sequences | Unit 2: Journey to the Moon | Reasoning | 36–39 |
| | | Unit 2: What's in the ring? | Reasoning | 40–43 |
| | Reasoning about numbers | Unit 4: Cinderella | Creative thinking | 60–63 |
| 9–10 | Place value, ordering, estimating, rounding | Unit 1: The Bear's breakfast | Information processing | 30–33 |
| | | Unit 2: Journey to the Moon | Reasoning | 36–39 |
| | | Unit 3: Hoopla | Enquiry | 50–53 |
| | | Unit 4: The Giant's footprint | Creative thinking | 56–59 |
| | | Unit 5: Aliens | Evaluation | 66–69 |
| | | Unit 6: Treasure hunt | Using and applying | 76–79 |
| | Understanding + and – | Unit 1: The Bear's breakfast | Information processing | 30–33 |
| | | Unit 2: Journey to the Moon | Reasoning | 36–39 |
| | | Unit 5: Aliens | Evaluation | 66–69 |
| | Mental calculation strategies (+ and –) | Unit 1: The Bear's breakfast | Information processing | 30–33 |
| | Money and 'real life' problems | Unit 1: The Bear's breakfast | Information processing | 30–33 |
| | | Unit 4: The Giant's footprint | Creative thinking | 56–59 |
| | | Unit 5: Aliens | Evaluation | 66–69 |
| | | Unit 6: Treasure hunt | Using and applying | 76–79 |
| | Making decisions | Unit 3: Hoopla | Enquiry | 50–53 |
| | | Unit 4: Cinderella | Creative thinking | 60–63 |
| | | Unit 5: Funny fish | Evaluation | 70–73 |
| | | Unit 6: The growing caterpillar | Using and applying | 84–87 |
| 11–13 | Measures, and time, including problems | Unit 4: The Giant's footprint | Creative thinking | 56–59 |
| | Handling data | Unit 2: What's in the ring? | Reasoning | 40–43 |
| | | Unit 5: Funny fish | Evaluation | 70–73 |
| | | Unit 6: The growing caterpillar | Using and applying | 84–87 |
| 14 | **Assess and review** | | | |

## Summer Term

| Unit | Unit topic | Thinking by Numbers | | |
|---|---|---|---|---|
| | | **Activity name** | **Thinking skill** | **Page no.** |
| 1 | Counting, properties of numbers and number sequences | Unit 2: Journey to the Moon | Reasoning | 36–39 |
| | | Unit 2: What's in the ring? | Reasoning | 40–43 |
| | | Unit 4: Cinderella | Creative thinking | 60–63 |
| 2–4 | Place value, ordering, estimating, rounding | Unit 1: The Bear's breakfast | Information processing | 30–33 |
| | | Unit 2: Journey to the Moon | Reasoning | 36–39 |
| | | Unit 6: Treasure hunt | Using and applying | 76–79 |
| | Understanding + and − | Unit 1: The Bear's breakfast | Information processing | 30–33 |
| | | Unit 2: Journey to the Moon | Reasoning | 36–39 |
| | | Unit 5: Aliens | Evaluation | 66-–69 |
| | Mental calculation strategies (+ and −) | Unit 1: The Bear's breakfast | Information processing | 30–33 |
| | Money and 'real life' problems | Unit 1: The Bear's breakfast | Information processing | 30–33 |
| | | Unit 4: The Giant's footprint | Creative thinking | 56–59 |
| | | Unit 5: Aliens | Evaluation | 66–69 |
| | | Unit 6: Treasure hunt | Using and applying | 76–79 |
| | | Unit 3: Hoopla | Enquiry | 50–53 |
| | | Unit 4: Cinderella | Creative thinking | 60–63 |
| | | Unit 5: Funny fish | Evaluation | 70–73 |
| | Making decisions, checking results | Unit 6: The growing caterpillar | Using and applying | 84–87 |
| 5–6 | Measures, including problems | Unit 4: The Giant's footprint | Creative thinking | 56–59 |
| | Shape and space Reasoning about shapes | Unit 1: Little Red Riding Hood | Information processing | 26–29 |
| | | Unit 2: What's in the ring? | Reasoning | 40–43 |
| | | Unit 3: How many shapes? | Enquiry | 46–49 |
| | | Unit 6: Beautiful butterflies | Using and applying | 80–83 |
| **7** | **Assess and review** | | | |
| 8 | Counting, properties of numbers and number sequences Reasoning about numbers | Unit 2: Journey to the Moon | Reasoning | 36–39 |
| | | Unit 2: What's in the ring? | Reasoning | 40–43 |
| | | Unit 4: Cinderella | Creative thinking | 60–63 |
| 9–11 | Place value, ordering, estimating | Unit 1: The Bear's breakfast | Information processing | 30–33 |
| | | Unit 2: Journey to the Moon | Reasoning | 36–39 |
| | | Unit 3: Hoopla | Enquiry | 50–53 |
| | | Unit 4: The Giant's footprint | Creative thinking | 56–59 |
| | | Unit 5: Aliens | Evaluation | 66–69 |
| | | Unit 6: Treasure hunt | Using and applying | 76–79 |
| | Understanding + and − | Unit 1: The Bear's breakfast | Information processing | 30–33 |
| | | Unit 2: Journey to the Moon | Reasoning | 36–39 |
| | | Unit 5: Aliens | Evaluation | 66–69 |
| | Mental calculation strategies (+ and −) | Unit 1: The Bear's breakfast | Information processing | 30–33 |
| | Money and 'real life' problems | Unit 1: The Bear's breakfast | Information processing | 30–33 |
| | | Unit 4: The Giant's footprint | Creative thinking | 56–59 |
| | | Unit 5: Aliens | Evaluation | 66–69 |
| | | Unit 6: Treasure hunt | Using and applying | 76–79 |
| | | Unit 3: Hoopla | Enquiry | 50–53 |
| | | Unit 4: Cinderella | Creative thinking | 60–63 |
| | | Unit 5: Funny fish | Evaluation | 70–73 |
| | Making decisions, checking results | Unit 6: The growing caterpillar | Using and applying | 84–87 |
| 12–13 | Measures, and time, including problems | Unit 4: The Giant's footprint | Creative thinking | 56–59 |
| | Handling data | Unit 2: What's in the ring? | Reasoning | 40–43 |
| | | Unit 5: Funny fish | Evaluation | 70–73 |
| | | Unit 6: The growing caterpillar | Using and applying | 84–87 |
| **14** | **Assess and review** | | | |

# Thinking by Numbers 1 and the NNS Framework

| Thinking skill | Unit | Activity name | Counting, properties of numbers and number sequences | Place value and ordering | Estimating | Understanding addition and subtraction | Rapid recall of addition and subtraction facts | Mental calculation strategies (+ and −) | Making decisions | Reasoning about numbers or shapes | Problems involving 'real life', money or measures | Organizing and using data | Measures | Shape and space |
|---|---|---|---|---|---|---|---|---|---|---|---|---|---|---|
| Information processing | 1 | Little Red Riding Hood | | | | | | | | ✓ | | | | ✓ |
| | 1 | The Bears' breakfast | | ✓ | | ✓ | | ✓ | | | ✓ | | | |
| Reasoning | 2 | Journey to the Moon | ✓ | ✓ | | ✓ | | | | | | | | |
| | 2 | What's in the ring? | ✓ | | | | | | | ✓ | | ✓ | | ✓ |
| Enquiry | 3 | How many shapes? | | | | | | | | ✓ | | | | ✓ |
| | 3 | Hoopla | | | ✓ | | ✓ | | ✓ | | | | | |
| Creative thinking | 4 | The Giant's footprint | ✓ | | ✓ | | | | | | ✓ | | ✓ | |
| | 4 | Cinderella | | | | ✓ | ✓ | | ✓ | | | | | |
| Evaluation | 5 | Aliens | | | ✓ | | | | ✓ | | ✓ | | | |
| | 5 | Funny fish | | | | | | | ✓ | | | ✓ | | |
| Using and applying thinking skills | 6 | Treasure hunt | | ✓ | | | ✓ | | | | ✓ | | | |
| | 6 | Beautiful butterflies | | | | | | | | ✓ | | | | ✓ |
| | 6 | The growing caterpillar | | | | | | | ✓ | | | ✓ | | |

# Thinking by Numbers 1 and the 5–14 Guidelines

| Thinking skill | Unit | Activity name | Problem-solving and Enquiry | Information Handling | Number, Money and Measurement | | | | | | | | | | | Shape, Position and Movement |
|---|---|---|---|---|---|---|---|---|---|---|---|---|---|---|---|---|
| | | | | | Range and Type of Numbers | Money | Add and Subtract | Multiply and Divide | Round Numbers | Fractions, Percentages and Ratio | Patterns and Sequences | Functions and Equations | Measure and Estimate | Time | Perimeter, Formulae and Scales | |
| Information processing | 1 | Little Red Riding Hood | ✓ | | | | | | | | | | | | | ✓ |
| | | The Bears' breakfast | ✓ | | ✓ | | ✓ | | | | ✓ | | | | | |
| Reasoning | 2 | Journey to the Moon | | ✓ | ✓ | | | | | | ✓ | | | | | |
| | | What's in the ring? | | ✓ | ✓ | | | | | | ✓ | | | | | ✓ |
| Enquiry | 3 | How many shapes? | ✓ | | | | | | | | | | | | | ✓ |
| | | Hoopla | ✓ | | | | ✓ | | ✓ | | | | ✓ | | | |
| Creative thinking | 4 | The Giant's footprint | ✓ | | | | | ✓ | ✓ | | | | ✓ | | | |
| | | Cinderella | ✓ | | | | | | | ✓ | | | | | | |
| Evaluation | 5 | Aliens | ✓ | ✓ | | | ✓ | | | | | | ✓ | | | |
| | | Funny fish | ✓ | | | | ✓ | ✓ | | | | | | | | |
| Using and applying thinking skills | 6 | Treasure hunt | ✓ | | | | | | | | ✓ | | | | | |
| | | Beautiful butterflies | | | | | | | | | | | | | | ✓ |
| | | The growing caterpillar | ✓ | ✓ | | | | | | | | | | | | |

# Glossary

**algorithm** a step by step procedure that, if followed exactly, will always yield a correct solution to a type of problem

**assessment for learning** an approach to **formative assessment** where the learner is encouraged to take responsibility for evaluating their own achievement of learning objectives. An aspect of **self-regulation**.

**Bloom's Taxonomy** a widely used instructional objectives model developed by the prominent educator Benjamin Bloom and colleagues in the 1950s. It categorizes the cognitive, affective and conative domains and includes a systematic list of thinking skills, in categories and sub-categories such as comprehension, application, analysis, synthesis, and evaluation. The last three are considered **higher-order** thinking skills.

**brain-based learning** a range of techniques and approaches to teaching and learning which take their inspiration from research into how the brain works to identify implications for teaching

**brainstorm** a technique for rapid production of ideas without critical examination, evaluation or elaboration

**bridging** a teaching strategy where explicit links are drawn from what has been learned to other related contexts to help **transfer**

**cognition** the mental operations involved in thinking; the biological/neurological processes of the brain that facilitate thought. Sometimes contrasted with affect or emotion and conation (wanting or willing).

**Community of Enquiry** the process of developing knowledge and understanding by participating in purposeful dialogue or collaborative discussion. Also the teaching technique used in Philosophy for Children with a class of pupils.

**concrete preparation** an introductory phase in some teaching thinking approaches where new words are introduced and learners become familiar with what the task is about

**constructivism** a view of learning in which learners are seen as building or developing their own understanding of how the world works from their experience and interaction with people around them

**creative thinking** producing new ideas or thoughts. Imaginative thinking that is aimed at producing outcomes that involve synthesis of ideas or lateral thinking; thinking that is not analytical or deductive, sometimes referred to as divergent thinking.

**critical thinking** a generic term for thinking skills used in the United States. The process of determining the authenticity, accuracy, or value of something; characterized by the ability to seek reasons and alternatives, perceive the complete situation, and change one's view based on evidence and reasoning. Sometimes also called analytical or convergent thinking. Often related to formal or informal logic and to reasoning.

**demonstrating** showing children how to do something, how to perform a skill or a technique, how to carry out a process, how to repeat and practise what they have been shown

**dialogue** shared enquiry between two or more people

**enquiry** a systematic or scientific process for answering questions and solving problems based on gathering evidence through observation, analysis and reflection

**enquiry learning** a teaching strategy designed to develop pupils learning through systematic gathering of observation and investigation

**enrichment** an approach to teaching thinking as separate discrete skills, usually as separate lessons using a particular programme or set of activities

**formative assessment** assessment which alters subsequent teaching and learning. This may involve teachers in using information gathered in lessons to alter what they do (see **mediation**) or it may also involve the learner through **assessment for learning**.

**graphic organizers** diagrams which help learners to organize information such as by comparing and contrasting using a grid of similarities and differences

**heuristics** general or widely applicable problem-solving strategies. Guidelines that generally direct attention, but that do not always produce a correct outcome (see **algorithm**).

**higher order thinking** evaluation, synthesis and analysis, the higher levels of **Bloom's Taxonomy**

**infusion** integrating thinking skills teaching into the regular curriculum or lessons; infused programs are commonly contrasted with **enrichment** programs, where separate or discrete skills are taught through lessons to promote thinking.

**mediation** a teaching strategy where the teacher intervenes and supports the development of pupils' understanding by **modelling** or by direct instruction to help them achieve something they could not do alone

**metacognition** the process of planning, assessing, and monitoring one's own thinking. Thinking about thinking in order to develop understanding or **self-regulation**.

**modelling** teaching children in a way that helps them to see the underlying structures, and to understand the embedded or supporting concepts and ideas

**multiple intelligences** the idea developed by Howard Gardner that IQ does not measure aspects of intelligence sufficiently and that people have strengths in different areas such as visual-spatial or musical as well as more traditionally assessed areas such as linguistic or logico-mathematical

**problem based learning** an approach using **problem solving** techniques where learners are set specific challenges through realistic or unstructured problems. Similar to **enquiry learning**, but with a particular goal or challenge which needs to be resolved

**problem solving** a general term which covers a diversity of problem types which make a range of demands on thinking. Some problems have unique solutions and can be tackled with predominantly convergent critical thinking, but many others are open-ended and demand both creative and critical thinking for their solution.

**reasoning** drawing conclusions or inferences from observations, facts, experiences: deductive inferring conclusions from premises; inductive: inferring a provisional conclusion or hypothesis from information

**self-regulation** the conscious use of mental strategies to improve thinking and learning, often aimed at particular learning goals

**seriation** sequencing or arranging objects, ideas or events in a particular order determined by a criterion

**Socratic questioning** an approach to questioning and discussion where answers to questions are pursued through dialogue

**thinking skills** 'thinking skills' and related terms are used to indicate a teaching approach which emphasizes the processes of thinking and learning that can be used in a range of contexts. The list of thinking skills in the English National Curriculum is similar to many such lists: information-processing, reasoning, enquiry, creative thinking and evaluation.

**transfer** the ability to apply an idea or a skill that has been learnt in one context and use it in a different context